ACCOUNTING

HISTORY AND THOUGHT

EDITED BY
RICHARD P. BRIEF
New York University

A GARLAND SERIES

AN ANALYSIS OF THE EARLY RECORD KEEPING IN THE DU PONT COMPANY 1800-1818

ROXANNE THERESE JOHNSON

GARLAND PUBLISHING, INC.
New York & London
1989

Library of Congress Cataloging-in-Publication Data

Johnson, Therese Roxanne.
An analysis of the early record keeping in the Du Pont Company 1800–1818 / Roxanne Therese Johnson.
p. cm. — (Accounting history and thought)
Includes bibliographical references.
ISBN 0-8240-3319-1 (alk. paper)
1. Gunpowder industry—United States—Accounting—History—19th century.
2. E. I. du Pont de Nemours & Company—Records and correspondence.
I. Title. II. Series.
HF5686.G93J64 1989
338.7'6234526—dc20   89-23350

Printed on acid-free 250-year-life paper

Manufactured in the United States of America

DEDICATION

I am indebted to William Schrader and Gerald Eggert of The Pennsylvania State University for their support and encouragement in the completion of this research. Their suggestions and advice provided invaluable insight into the materials and their meaning. In addition, the enthusiasm of Steve Jablonsky and Roland Pellegrin, also of The Pennsylvania State University, was greatly appreciated.

My heartfelt appreciation goes to my family and friends and I especially want to acknowledge my parents who have always had confidence in my abilities.

I would also like to thank the Hagley Library's Manuscripts and Archives, Pictorial and Imprints Departments.

TABLE OF CONTENTS

LIST OF ILLUSTRATIONS

ILLUSTRATION 1. Examples of the record books used by the DuPont Company record keepers at the beginning of the nineteenth century. Accession 500, Series I, 1800–1818. Photograph courtesy of Hagley Museum and Library.

PREFACE

I first learned about the DuPont Company's origins in gunpowder manufacture during a tour of a DuPont family legacy, the Hagley Museum. The museum, located at the site of the original DuPont gunpowder works, depicted manufacturing life along the Brandywine River in the nineteenth century using scrupulously reproduced exhibits and renovated buildings.

The first step in this research was a literature review of prior and current research on these materials. Remarkably, little work on this period exists. In an unpublished Master's Thesis on the subject of the DuPont bookkeeping practices of this early period, Edwards dealt with the bookkeeping system from an historical perspective as opposed to analyzing in detail the significance of the trends, techniques and developments evident in the record keeping.[1] This work is a valuable resource for individuals with little or no knowledge of bookkeeping but it fails to address the significance of the records in terms of the history of accounting. The opportunity for uncharted research on these records was, surprisingly, still available.

This volume examines the earliest primary materials of the DuPont Company based on the extant records of the period under study, 1800 to 1818, which are preserved at the Hagley Library, another DuPont legacy.[2] These were the formative years of the company's record keeping since systematic procedures were essentially in place by August 1818. The integrity of the materials used in this study was verified with the assistance of the Hagley Library's bindery expert, Mr. David Moore.

Certain secondary materials to support my research also were used.[3] The secondary source of most value was an eleven-volume set of family letters entitled <u>Life of Eleuthère Irenée du Pont from Contemporary Correspondence, 1778-1834</u>, translated into English by Bessie Gardner DuPont.[4] Another valuable source was a <u>Guide to Manuscripts in the Eleutherian Mills Historical Library</u>.[5] This volume helped me to identify other materials which might have some bearing on my research. In addition I accessed late eighteenth-century contemporary works on record keeping to enhance my understanding of the bookkeeping practices of the time.[6]

The hypothesis of this study is that "usefulness" was the primary criterion which prompted the changes that occurred in the bookkeeping techniques used by the DuPont Company during the period under study. Even when the bookkeeping techniques used evidenced the education, background and experience of particular record keepers, the criterion which made the practice of double-entry bookkeeping the system of choice since the Middle Ages still applied.

Notes--Preface

1   Nina Lorraine Edwards, "The Bookkeeping Records and Methods of E. I. DuPont de Nemours and Company, 1801-1834" (M.A. thesis, University of Western Ontario, 1966).

2   The account books for the period are catalogued within the Hagley Library's Accession 500, entitled "Records of E. I. Du Pont de Nemours & Co., 1801-1902." The examination of the account books was facilitated by a guide to the Library's company manuscripts, the "Schedule of the Records of E. I. DuPont de Nemours & Co., 1802-1902," prepared by the Manuscripts Department of the Eleutherian Mills Historical Library (now Hagley Library) and dated January 1, 1965.

3   The general works referenced are indicated below, although other materials were accessed and cited as necessary.

4   Bessie Gardner DuPont, _Life of Eleuthère Irenée duPont from Contemporary Correspondence, 1778-1834_, trans. and ed. B. G. DuPont, 12 vols. (Newark, Delaware: University of Delaware Press, 1923-1926) (hereafter cited as DuPont, _Life_). The translations in these volumes are generally considered accurate in terms of content. Some specific translations are glaringly inaccurate, however, and where these have resulted in misleading information I have verified the true meaning of the terms in question. For the purposes of this research most general descriptions are taken verbatim from B. G. DuPont's work unless otherwise indicated. In some instances, B. G. DuPont only partially translated documents. Where applicable, this researcher has recorded the appropriate translation.

5   John Beverley Riggs, _A Guide to the Manuscripts in the Eleutherian Mills Historical Library_ (Greenville, Delaware: Eleutherian Mills Historical Library, 1970) (hereafter cited as Riggs, _Guide_).

6   I did not consider an exhaustive review of contemporary works on bookkeeping necessary for this analysis, but rather concentrated on a limited number of works with many published editions, specifically

Thomas Dilworth, <u>The Young Book-keeper's Assistant</u>,
12th ed. (Philadelphia: Benjamin Johnson, 1794)
(hereafter cited as Dilworth, <u>Assistant</u>) and John
Mair, <u>Book-keeping Moderniz'd</u>, 6th ed. (Edinburgh:
Bell & Bradfute, 1793; reprint ed., New York: Arno
Press, 1978) (hereafter cited as Mair, <u>Moderniz'd</u>).
Mair also had numerous previous editions under another
title. I considered these works representative of
techniques practiced at this time. Direct quotations
from these works have been altered as necessary in
accordance with current spelling practice for purposes
of clarity.

INTRODUCTION

The purpose of this research is to analyze the record keeping techniques and developments evidenced in the DuPont Company account books during the early nineteenth century. Essential to such an analysis is a thorough understanding of the history of the company for which such records were maintained, E. I. DuPont de Nemours & Company. As it usually is in family-dominated firms, the history of the company is inextricably tied to the history of the family. The DuPont family emigrated to the United States from France in 1799. The patriarch of the family, Pierre Samuel DuPont de Nemours, engineered the emigration.[1]

Pierre Samuel DuPont was born December 14, 1739, in Paris, to Samuel DuPont and Anne Alexandrine de Montchanin. Pierre enjoyed only a brief formal education during his early years before Samuel, a watchmaker, forced his son to learn his trade. Although Pierre eventually mastered his father's occupation, he never liked it and refused to make it his life's work.

Instead, Pierre chose to pursue his own course with the hope, eventually, of securing a responsible position in government. Early in life he developed a number of personal characteristics which served him well in this effort. He became an extremely self-motivated individual with diverse interests and a great deal of curiosity. He combined these attributes with almost obstinate self-confidence and fortitude that occasionally resulted in stubborn, unyielding resolve. Nonetheless he enjoyed and welcomed new experiences and opportunities, and was remarkably undeterred by events of the past. He therefore

1

eagerly and capably continued his education on his own despite the lack of additional formal training.

In his pursuit of knowledge and a government position, Pierre developed a moderate approach to issues that interested him. His ideas were compatible with those of the Physiocrats, moderate reformers who believed that land was the only source of wealth and agriculture the basis of a healthy society. This was the first organized group to recognize his potential. He attracted the attention of leading members of the movement, specifically Dr. François Quesnay, who encouraged him to do some work on another important issue, the free importation and exportation of grain. His work in this area favorably impressed his supporters and helped to secure his reputation as a dedicated and tireless worker.

From 1763 to 1774 Pierre had no stable occupation or income. He took many different jobs during this period, relying on whatever sporadic employment opportunities he could find to earn a living. He eventually assumed a government position as Inspector General of Commerce in September 1774 at the instigation of the Controller-General, Anne Robert Jacques Turgot, another renowned moderate reformer. Pierre retained his position despite dramatic changes in the French political climate during the ensuing years leading to the most turbulent period in French political history, the French Revolution.

Consequently on May 4, 1789, the first meeting since 1615 of the Estates General, a representative body summoned only in times of great trial, convened in Paris. This gathering soon converted itself into the Constituent Assembly. Pierre attended the meeting as one of two delegates elected to represent Nemours. He became a well-known figure in the Assembly because of his detailed and exhaustive work on several committees. He had played a relatively insignificant part in earlier events but he was too outspoken to remain in the background forever. Nonetheless, politically and economically he maintained his moderate perspective and refused to follow the more extreme elements within the Assembly. This attitude brought him into conflict with the more adamant reformers of the period although he still managed to retain the respect of many of the delegates at the Assembly and therefore continued to contribute to and participate in events.

Over time public opinion and prudence forced Pierre to relinquish all his government subsidies and positions, although any income he had received for his services had been sporadic and unreliable. In an effort to augment his diminished resources he started a print shop not long before the Constituent Assembly disbanded in 1791.[2] He had learned the printing profession in one of his former positions as editor of a journal. In his capacity as head of the printing office he edited and published the writings of others, and still contributed his own work as well. Indeed, he used this forum to continue to criticize and attack the extremist factions which dominated the government.

In the process Pierre particularly alienated the Jacobins who assumed control of the government in July 1793. False accusations brought before the dominant Jacobin group concerned Pierre's alleged involvement in active opposition to the revolutionary cause. Although he had limited his opposition to demands for moderation this group still considered him a threat. Finally in July 1794 the Committee of Public Safety, by now dominated by Robespierre, ordered his arrest. He was taken to the prison La Force in Paris and was scheduled to die. The fall of Robespierre on July 27, 1794, spared his life.

After this, Pierre essentially withdrew from the political mainstream and concentrated his efforts on the print shop. He resumed his political career in late

October 1795 when the government of the Directory took control. Pierre sat in one of the two chambers of the new legislature, the Council of Elders. Executive power rested with the five-member Directory selected by the Council of Elders from a list submitted by the other legislative chamber, the Council of the Five Hundred. Pierre's moderate philosophy again placed him at odds with the extremist elements prevailing in the government.

In 1797 France experienced further political upheaval and unrest. Country-wide elections beginning in April of that year resulted in bitter and violent contests and a massive turnover in the government. On September 4, 1797, three of the Directors who believed that the royalists would attempt to take advantage of the confusion and regain political control initiated a coup d'état. These Directors next moved to arrest perceived enemies among the rest of the Directory and the two Councils. Pierre and one of his sons were among those arrested, but both were released within twenty-four hours. Pierre's moderate stance had again placed him in jeopardy, however. Pierre officially resigned from the Council of Elders soon after this incident, on September 13, 1797.

Domestically as well as politically Pierre faced many trials, tribulations and successes. On January 28, 1766, he married Nicole-Charlotte-Marie-Louise Le Dée de Raucourt and started a family despite his straitened economic circumstances during this period. His first son, Victor, was born in October 1767. His wife gave birth to a third son (the second had died) named Eleuthère Irenée in June 1771. Thirteen years later, in 1784, she suddenly died. Up to this point, she had essentially complete responsibility for the domestic side of Pierre's life. Following her death, Pierre had to assume the management of the family estate, Bois des Fossés, and responsibility for the two children. He did not remarry until September 26,

4

1795. His second wife was Françoise Poivre, the widow of one of his friends.

Pierre's two sons could not have been more unlike each other. Victor, the elder son, was something of a dandy. His early letters to his brother were filled with missives and requests concerning his attire. They were also filled with wit, satire and criticism, usually aimed at his brother's more earnest existence and occasional remissness.[3] These qualities may have been fostered in large measure because Victor, seventeen when his mother died, did not have to endure the brunt of his father's interminable direction. Pierre strongly disapproved of Victor's attitudes, interests and evident lack of ambition, and was severely disappointed by his elder son.[4]

Despite his father's impression of him, Victor eventually decided to pursue a career in foreign service, a profession suiting his heretofore unappreciated qualities. He launched his career in 1788 when he found an unpaid position with the French envoy to the United States. This apparently resulted in two diplomatic missions to the new country, the second ending in the summer of 1793. In early 1794 Victor married Josephine Gabrielle La Fite de Pelleport. Following the marriage, he secured a position with the Bureau of Foreign Relations and later returned to America as the first secretary of the French legation. Later, he served as consul at Charleston, South Carolina. Sometime in 1798 he learned of a vacancy in the consular seat in Philadelphia, which he obtained. Unfortunately, relations between France and the United States were strained and he could not fill his post. Nonetheless, he still planned to continue in foreign service and appeared to have a successful career ahead of him.

Pierre's second son, Eleuthère Irenée (hereafter E. I.), was very different from his brother and did not seem to have Victor's frivolous nature. Only thirteen when

5

his mother died, he grew up under his father's rather taxing supervision. Pierre's letters to his son are filled with exhortations to work hard and long, maintain an even temperament, and remain obedient to his father's wishes.[5] Pierre essentially groomed E. I. to be the kind of son he wanted--responsible, resolute, serious and hardworking, the exact opposite of what Pierre perceived Victor to be.

E. I. expressed an early interest in the manufacture of gunpowder.[6] This apparently influenced, or possibly predetermined, his career path. He benefited from his father's friendship with a highly respected individual, Antoine Laurent Lavoisier, the world-renowned chemist. Pierre met Lavoisier when both were members of a special government committee formed to investigate agricultural problems. Lavoisier managed the government's gunpowder ministry, the Administration of Powder and Saltpetre, located at the Paris Arsenal. Sometime in late 1788 or early 1789 E. I. became a management apprentice at the Administration and enjoyed a widely diverse, scientifically oriented education both at the Arsenal and at a gunpowder mill located outside of Paris at Essonne. During his course of study, he specifically learned how to manage a gunpowder manufactory.

While still at the Administration, E. I. married Sophie Dalmas on November 26, 1791, despite formidable initial opposition from his father. His own arguments in favor of the marriage focused on his position at the Administration and the likelihood of advancement. He finally won Pierre over, but unfortunately he lost his position shortly thereafter when Lavoisier resigned from the Administration and left the arsenal in 1792.[7]

E. I. subsequently joined his father in the management of the printing office. He became the primary manager of the firm whenever Pierre's duties, responsibilities or occasional need to flee Paris as a fugitive left him in charge. During this period, therefore, he also received valuable management training. When Pierre was arrested the second time, E. I. was also taken into custody.

It was not clear exactly when Pierre formulated his intention to leave France. His arrests certainly served to warn him that he, as well as members of his immediate family, would not be safe as long as they remained in France. Following his resignation from the Council of Elders, Pierre and his family spent the remainder of their time in France preparing for the journey to America.[8]

As mentioned, Pierre occasionally overstepped his capabilities, took unnecessary chances and tended to leap into situations without adequate preparation or forethought. Nowhere was this more evident than when he engineered the family's emigration to America. For the family to emigrate, they had to have some means of making a living once they arrived in America since the DuPonts were not wealthy. To accomplish this, Pierre devised a plan for a family business that he hoped to finance before he left France. Perhaps because of his physiocratic background, he became interested in a business which in some way involved land management and development.

In a 1797 prospectus directed at potential investors Pierre outlined an ambitious plan for the firm, DuPont de Nemours Father, Sons & Company. He specifically proposed that the firm engage in land speculation, management of an agricultural community and associated commercial enterprises, and a commission business. Pierre estimated that the firm's capitalization requirements were such that they would have to sell "at least two hundred shares of ten thousand francs each and . . . not . . . more than four hundred."[9] The principals of the firm were identified in the prospectus as "Citizen du Pont de Nemours, . . . Citizen Bureaux de Pusy, an expert engineer, . . . Citizen du Pont, junior, French consul at Philadelphia; and his brother Irénée du Pont, a business man."[10]

Pierre's two sons participated in his new venture, although their reasons for doing so differed considerably. Victor apparently had no alternative other than to become a principal in the new firm. He returned from America in July 1798 to find himself already named in the new firm's prospectus. He was unable to extricate himself from the proceedings, largely because of his sense of family responsibility, although it appears that he really wanted to remain in the foreign service.[11]

E. I., on the other hand, continued to manage the print shop. This took all of his time, and despite his efforts the business suffered as conditions worsened in France. E. I.'s letters to his wife during this time were full of remorse and despair at their continued forced separation as he toiled in Paris and she at the family estate. The situation worsened as the mails became increasingly less certain and timely.[12] For these reasons E. I., as well as his family, probably welcomed the proposed move.

One point to keep in mind when considering the decisions and expectations of the DuPonts at this time is that Pierre never sought either guidance or background information for his plans from sources more familiar with conditions in America. Therefore an obvious source, his son Victor, was not approached to gather information or provide advice, and only discovered the situation and his part in it upon his return from America. Nor did he consult with another individual who could have been of even greater service in advising him, his old friend, Thomas Jefferson. Pierre's willfulness meant the family embarked on this enterprise without adequate knowledge of conditions and opportunities in the United States.

Two years after the 1797 prospectus, when Pierre was finally ready to move to America, he had received only about one-fourth of the amount he had initially calculated as the minimum needed to begin operations. The firm Du Pont de Nemours Father, Sons & Company was therefore seriously under-capitalized from the very beginning.[13] Nonetheless, Pierre and the family members travelling with him arrived in the United States at the turn of the century.[14] Despite the firm's straitened circumstances, Pierre still believed it would be successful.

At some point Pierre determined that he should not advertise the land speculation aspect of the business immediately in case such information limited the firm's opportunities.[15] He planned to use this subterfuge only until he had investigated the land speculation opportunities which existed. An undated circular with the address, "New-York," therefore detailed the extent of the proposed commission business only.[16]

It was just as well, since within a year Pierre himself recognized the inadvisability of his original intentions for the firm, and even expressed thanks that he had not been able to invest in the schemes proposed to him upon his arrival.[17] He also faced the fact that the fledgling commission business had not attracted the clients in America he had expected.[18] In late 1800, therefore, Pierre outlined eight plans which covered new opportunities he hoped the firm would be able to exploit. He detailed his two sons to carry his ideas back to France.[19]

Victor's objectives when he returned to France were to acquaint the existing investors with the change in focus of the venture, to give the reasons for that change, and to reassure them as to the viability of the firm. He was also to seek additional investors, if possible. In addition, although he was not specifically charged with activation of a particular plan, he began negotiations which ultimately involved the firm in a scheme to supply French ships stationed in Santo Domingo.

Of the eight plans, Pierre gave E. I. responsibility for the eighth--to establish a gunpowder manufactory. E. I. had spent considerable time convincing his father to support his project. In his favor, his detailed analysis of the market potential for high quality gunpowder was accurate, as both domestic and imported powder were of low quality and poorly manufactured.[20] Pierre in his turn became enthusiastic about the project and in a letter to his chief backer stated that E. I.'s "skill in this art, the ignorance of it in America, the needs of Government, those of the Country and even of the Spanish Indies, gives us not hope but a positive certainty of great profits."[21] Nonetheless, these expectations and plans were somewhat optimistic and precipitous in many ways.

E. I. estimated that it would cost $36,000 to build, man and begin operating the kind of gunpowder manufactory he envisioned.[22] He expected this amount to buy the optimal location and buildings he considered essential for the kind of high quality operation he planned.[23] In addition to the lands and buildings, he also needed personnel, raw materials and equipment to successfully begin operations. He went to France to raise the necessary funds to finance his undertaking, and procure the personnel and equipment he needed to operate his manufactory. E. I. renewed his acquaintance with the members of the French Gunpowder Ministry in Essonne, and updated his knowledge of manufacturing techniques and equipment. He also purchased advanced equipment that made his firm capable of state of the art gunpowder manufacturing.[24] In addition, while he was in France, the "Deed of Association" for E. I. DuPont de Nemours & Company which identified the particulars of the gunpowder manufactory, the shareholders, and operational expectations, was drafted in Paris on April 21,

Estimation des depenses necessaires pour l'établissement
de la manufacture de Poudre

à payer dans la 1ere année.

achat du Terrein ~~environs~~ de 6000 dont un tiers content	2000.
machines construites en france	4000.
avances à faire aux ouvriers venant de france	1000.
Logement des ouvriers	1000.
Raffinerie	1500.
Moulin à meules	2000.
Composition et magazin à charbon	400.
moulin à Pilons	2000.
moulin à grainer	1200.
grainoir et Presses	500.
Lissoir	2000.
Sechoir	400.
Epoustoir et enfonçage	500.
Magazin	1000.
Cloture	1500.
Maison du directeur et Barn	3000.
	23000

à payer dans la ~~seconde année et les~~
~~suivantes~~ — fonds libres pour
l'achat des matières premieres, la construction
d'un second moulin à Pilons, et les payemens
successifs du terrein —— 13000

36000,

ILLUSTRATION 2. The detailed list of the proposed costs of the manufactory, in E. I.'s handwriting. Longwood Manuscripts, Group 3, #2367. Photograph courtesy of Hagley Museum and Library.

1801.[25] At this point, the manufactory existed on paper
only.

After E. I. convinced his father to support the
gunpowder manufactory, Pierre made little contribution to
the enterprise. In fact, on May 1, 1802, the family
officially established the firm DuPont de Nemours Father,
Sons & Company in Paris, and ceased operations in the
United States. The American commission business, such as
it was, was turned over to a firm managed by Victor and
called Victor DuPont de Nemours & Company.[26] The latter
firm soon became a casualty of Victor's efforts to arrange
to finance French ships in the Caribbean on behalf of the
French government. The arrangement he engineered with the
French had never resulted in a signed agreement, and the
results were ultimately disastrous for his firm. In
addition, the firm DuPont de Nemours Father, Sons &
Company was dissolved on May 20, 1811.[27] As events
developed, E. I. consequently became the Director of the
only successful DuPont family endeavor, E. I. DuPont de
Nemours & Company, able to guarantee the future prosperity
of the family.

Eventually the company would come to dominate the
gunpowder industry; but in the early years of the
manufactory's existence, E. I. found himself in grave
financial trouble over and over again. The record books
used to document the early operations of the gunpowder
manufactory chronicle this struggle. The fact that the
firm survived this period in American economic history at
all is important. The record keeping practices used,
although they cannot be considered generally applicable to
other firms of the period, do shed light on the motivating
factors which caused changes in the procedures used.

Notes--Introduction

1   The following account of the history of the family, except where otherwise indicated, is based on the excellent account by Ambrose Saricks, _Pierre Samuel Du Pont de Nemours_ (Lawrence, Kansas: The University of Kansas Press, 1965), pp. 1-272 (hereafter cited as Saricks, _de Nemours_). The form DuPont is used for both the family and company name for purposes of consistency, although other forms exist.

2   He published a prospectus on the printing office in June 1791. "Printing Office of du Pont Deputy from Nemours to the National Assembly," June 8, 1791. DuPont, _Life_, I:141.

3   For examples of such letters, see Victor DuPont to E. I. DuPont, December 18 & 26, 1786 and January 3 & 12, 1787. DuPont, _Life_, I:64-74.

4   This was made clear in a letter E. I. received from Pierre. P. S. DuPont to E. I. DuPont, July 19, 1786. DuPont, _Life_, I:45-6.

5   For examples of such letters, see P. S. DuPont to E. I. DuPont, July 19 & 25, 1786 and August 15, 1786. DuPont, _Life_, I:45-7, 54-7.

6   Pierre referred to this interest in a letter to his son. P. S. DuPont to E. I. DuPont, March 22, 1785. DuPont, _Life_, I:33.

7   For details of Lavoisier's career with the Administration see Sidney J. French, _Torch and Crucible_ (Princeton: Princeton University Press, 1941); Douglas McKie, _Antoine Lavoisier; Scientist, Economist, Social Reformer_ (New York: Henry Schuman, 1952); and Henry Guerlac, _Antoine-Laurent Lavoisier_ (New York: Charles Scribner's Sons, 1975).

8   Although references to the proposed move appear in family letters as early as September 1797, the DuPont family finally emigrated to the United States in late 1799. E. I. DuPont to Sophie DuPont, September 27, 1797. DuPont, _Life_, IV:68-71.

13

9   The title of the firm has been translated from the French, DuPont de Nemours Père, Fils et Compagnie. In a prospectus endorsed "Project of a Rural Society--1797," Pierre outlined his plans and expectations for the enterprise. "Outline of a Plan for an Agriculture and Commercial Establishment in the United States of America." DuPont, Life, IV:86-100.

10   The three DuPonts were Pierre, Victor and E. I., respectively. DuPont, Life, IV:87-88. Citizen Bureaux de Pusy was Pierre's second wife's son-in-law.

11   B. G. DuPont translated a section of Mrs. Victor DuPont's reminiscences in which Victor's wife discussed her husband's preferences. "Our Transplantation to America." DuPont, Life, IV:118-121.

12   Letters from E. I. to his wife chronicled this situation. For examples, see E. I. DuPont to Sophie DuPont and vice versa. DuPont, Life, III, IV, V.

13   In an undated manuscript drafted prior to the family's emigration Pierre discussed the "Present Status of our Project." At this point he indicated that "We have had subscriptions or promises either verbal or written for more than three hundred shares sold or reserved." He described the capital investment for the firm in detail in four "chapters." The first Chapter was called "Payments already made" and included investment in terms of land and money or bills of exchange. Chapter Two was called "Payments that will surely be made." Chapter Three was called "Still to be hoped for, but less sure," and Chapter Four contained "Possibilities." "Present Status of our Project," undated. DuPont, Life, V:99-109. Unfortunately little of this was actually realized, and despite Pierre's expectations the firm received no other promises or active investment beyond that received prior to the family's removal to America. Immediately prior to the move E. I. disposed of the print shop and its equipment. E. I. DuPont to Sophie DuPont, May 21, 1799. DuPont, Life, IV:341. Nonetheless, the family had few resources to draw on. Further, what investment they did have suffered a blow when the bank in which some of the funds were deposited failed.

14   The family actually arrived in two groups. The first group consisted of Pierre's wife and her son-in-law, Bureaux de Pusy, who were sent earlier to arrange for

living quarters and make other necessary preparations for the arrival of the rest of the family.

15 The fifth chapter of the undated manuscript "Present Status of Our Project" was called "What is to be done." In this chapter Pierre acknowledged "it is better therefore that we should announce ourselves as doing only a shipping business on commission, and that we should offer to refer purchases of land to our friends in Europe, so that we will have plenty of time for investigation. While we are investigating and deciding on our definite plans, we should increase our capital by doing a commission business which will require but a small part of our capital and give us an immediate profit." "Present Status of Our Project," undated. DuPont, Life, V:107-108.

16 The circular stipulated that the firm would "do business on commission," and would arrange for "paying or receiving accounts, storing merchandise, exchange and proper economy." Other pursuits would include collecting and forwarding "dividends or payments due from Congress on its various loans," arranging "investments or reinvestments that Europeans may want to make in this country," enforcing "the payment of sums due to Europeans, either by the State or by individuals," and directing "the administration of estates owned in this country by Europeans." Untitled. DuPont, Life, V:117-119.

17 P. S. DuPont to Jacques Bidermann, December 1, 1800. DuPont, Life, V:167. It is not clear exactly when Pierre realized the inadvisability of his original intentions for the firm, at least for the immediate future. It is possible that the aforementioned circular veiled his intentions to investigate the opportunities in the New World before advertising the true nature of the firm. In this case the commission business served, as initially intended, to occupy and finance the firm only until the initial objectives were met. It is also possible that Pierre had already begun to revise his original intentions with respect to the firm because despite his expectations land speculation in the United States had reached its peak and the opportunities Pierre originally envisioned for the firm were essentially nonexistent. Thomas Jefferson, for instance, when he finally learned the true nature of the DuPont enterprise, advised strongly against the effort. Pierre apparently did participate is a minor way in some land improvement projects, however. P. S. DuPont to Dr. DuPont at Rotterdam. DuPont, Life, V:142.

18   This may have been because of a fear and distrust of foreigners by Americans.

19   Pierre outlined his eight plans in a letter to his principal backer. P. S. DuPont to Jacques Bidermann, December 1, 1800. DuPont, Life, V:163-196.

20   E. I. outlined his expectations in an undated critique in which he noted that "there are already in the United States two or three plants that make bad powder at great expense and that nevertheless do a good business." "On the Manufacture of War and Sporting Powder in the United States," undated. DuPont, Life, V:199.

21   P. S. DuPont to Jacques Bidermann, December 1, 1800. DuPont, Life, V:191-192.

22   See appendix B for a detailed list of these proposed costs.

23   E. I. detailed the location and buildings he needed in "The Locations and Constructions Necessary for Manufacture of Gunpowder," undated. DuPont, Life, V:206-212.

24   In a copy of a record of E. I.'s expenses in Paris are entries for equipment purchased from the Arsenal. DuPont, Life, V:213. Bottée, Superintendent of Powder and Saltpetre to E. I. DuPont, April 15, 1802. DuPont, Life, VI:15-18.

25   A translation of the original Deed of Association can be found in appendix A.

26   An untitled notice dated May 1, 1802 detailed these events. DuPont, Life, VI:41-2.

27   Accession 360. Auguste de Staël to E. I. DuPont, March 1, 1813. DuPont, Life, IX:90.

Chapter One

THE BOOKS AND THE BOOKKEEPERS

Modern day business organizations are made up of
people--owners, managers, employees and, peripherally but
essentially, creditors. The same personnel requirements
existed for nineteenth-century enterprises. Then, as now,
all of these people rightfully demanded information
concerning some aspect of the firm's operations. The
employees needed to know their outstanding wages, or the
amount of any debts they owed to the firm; creditors wanted
assurances as to the viability of the firm; managers needed
to track the day-to-day operations of the firm; and the
owners required information on their investments. The
requirement for specific information placed demands on an
enterprise for a systematic documentation of events. Since
no designated authority in the nineteenth century mandated
the record keeping procedures to be followed by a
particular firm, responses to these information demands
depended entirely upon the individuals responsible for the
documentation process, their perceptions of the needs of
the organization, and how they chose to translate those
perceived needs into pragmatic record keeping. In turn,
the sophistication of the documentation process depended
upon the background of all of the individuals actually
involved in the determination of specific record keeping
requirements.

This group included the designated bookkeeper for the
firm, if there was one. No educational or experiential
prerequisites existed for the position, however. Thus, in
addition to or in the absence of an experienced bookkeeper
other decision makers, probably the owners and/or the

17

managers of the firm, influenced, dictated, or even assumed
the record keeping responsibilities for the firm. They may
or may not have received formal education or experience in
the practice of bookkeeping, although familiarity with
systematic record keeping techniques may have been a
routine, albeit informal, aspect of a nineteenth-century
individual's background. In any event, there were numerous
contemporary books on the practice of bookkeeping which
would be of assistance even to an individual with no
training at all.[1] No matter how much bookkeeping
experience or training these record keepers had, however,
they needed to employ record keeping techniques which were
adapted to the specific requirements of the firm under
consideration.

Many different individuals served the DuPont Company
in the capacity of record keeper during the period in
question. These bookkeepers documented the daily events
which affected the company in numerous account books.[2]
Initially, the account books were merely rudimentary
listings of events because the earliest record keeper was
apparently not fully trained as a bookkeeper. Over time,
however, the record keeping became more sophisticated as an
individual with more knowledge and experience assumed the
responsibilities of the primary record keeper.[3]

Article Four of the original Deed of Association which
formed the DuPont Company identified the namesake of the
firm, Eleuthère Irenée DuPont, as the primary manager for
the enterprise. In addition, Article Eleven stipulated
that the "Director of the manufactory will adopt for his
accounts the principles adopted by the Administration of
Powder & Saltpetre of France."[4] E. I.'s background and
experience in the making of gunpowder established him as
the obvious choice to direct the operations of the firm,
but his influence on the overall record keeping was
difficult to establish. Even though his position as

18

Director made him the primary decision maker for the company, his abilities in the practice of bookkeeping were suspect.[5] Neither do his contributions to the record keeping of the firm provide any evidence of his capabilities in this area. He did not function as the primary record keeper of the firm or set up the original books, but only maintained one of the supporting account books.[6]

Instead, the primary responsibility for the earliest DuPont Company record keeping fell initially to Peter Bauduy. He was born Pierre de Bauduy de Bellevue in France in June 1769. He left school at seventeen, joined the French army, and remained in France until his father summoned him to the family home, by now in Santo Domingo, when the French Revolution threatened. He remained in Santo Domingo until the slave rebellion there forced him to flee the troubled island in September 1791. In October he and his wife landed in Philadelphia and soon settled in Wilmington, Delaware with other French emigrants.[7]

Peter Bauduy was an ambitious and enterprising individual who made many contacts in the area. He also had some capital at his disposal for investment purposes. His ability to arrange credit for the new DuPont Company along with his own investment capital attracted E. I.'s attention when he was seeking funding for his firm. Despite initial differences Peter Bauduy became a partner in the DuPont Company on August 25, 1802.[8] Bauduy apparently assumed the record keeping responsibilities at about this time.[9] From all indications he was not an experienced bookkeeper, and his efforts by themselves did not result in adequate record keeping for the firm. He nonetheless continued as principal record keeper until February 11, 1806, when a far more experienced individual, Raphael Duplanty, became involved in the record keeping.[10] Peter Bauduy remained a partner in the firm, however, and continued his association

19

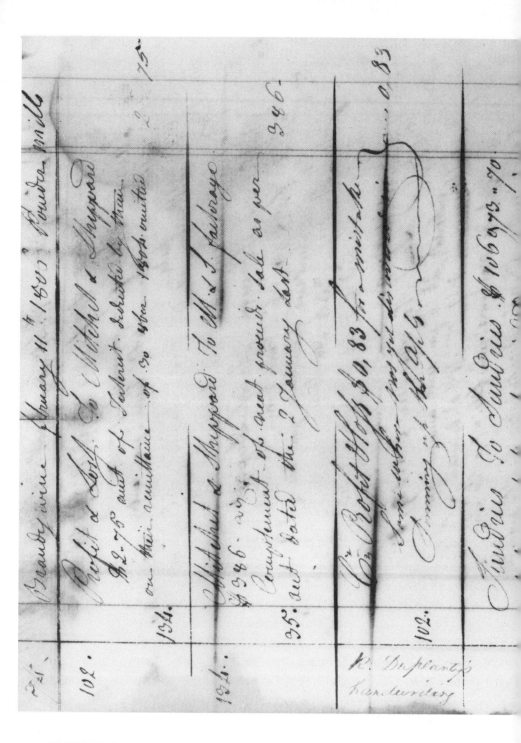

ILLUSTRATION 3. An example of the handwriting of Peter Bauduy and Raphael Duplanty, including a correcting entry by the latter, dated February 11, 1806. Accession 500, Journal, #877, folio 74. Photograph courtesy of Hagley Museum and Library.

with the DuPont Company. He also continued to perform some of the record keeping duties.[11]

Raphael Defredat Duplanty was born in 1776 in Brittany. Not much is known of his early life, although he did spend some time in England as a refugee from the French Revolution after his father and two brothers were guillotined and he was himself wounded and imprisoned. From there, he made his way to the West Indies, possibly to Santo Domingo. In 1803 he arrived in New York, and was introduced to Victor DuPont by the Consul there. He became involved in business in New York and at least part of the time worked for Victor.[12] In late 1805 DuPlanty was sent by Victor to Wilmington and eventually agreed to assume responsibility for the record keeping for the DuPont Company.[13] He seems to have spent some time reviewing the records before he assumed control.[14]

When Raphael Duplanty assumed primary responsibility for the record keeping functions of the DuPont Company on February 11, 1806, the accounts were in a sorry muddle. Peter Bauduy had only initiated one principal account book.[15] This first volume included a rough day-to-day record of events affecting the firm beginning with the shareholder investments.[16] Not only were the earliest entries in the volume documented long after they occurred, but Bauduy never carried the information he documented to any other record book for purposes of determining the firm's economic position.[17] Therefore, Raphael Duplanty essentially initiated a revolution in the bookkeeping techniques employed by the firm when he assumed control of the record keeping. He had a readily apparent advantage over Bauduy in that he had obviously had some experience and possibly education in the practice of the best record keeping method available, double-entry bookkeeping.

Usefulness as a functional requirement historically supported the practice of double-entry bookkeeping first documented in 1494 by Lucas Paciolo as part of a published work on algebra.[18] The methods Paciolo described still constituted the basis of the bookkeeping systems evidenced in nineteenth-century literature. The process had evolved somewhat, however, simply because of the changes in business practice and construct over the intervening years. The growth of the "extended firm" using distant agents to distribute goods geographically had introduced some new techniques into the literature of the day, although the basics of Paciolo's system still prevailed. Thus, in 1800, the most popular authors in contemporary bookkeeping literature continued to conform to Paciolo's basic tenets.[19]

Double-entry bookkeeping is a stylized, transactions-oriented method of documenting the day-to-day operations of a firm. Under this system, a chart of accounts was identified consisting of each individual or firm with whom the company did business, all items of value owned by the company and other elements as necessary. The first type of accounts, called "personal accounts," were identified with a person's or firm's name.[20] The second, called "real accounts," included all raw materials inventories, real estate and finished product inventories. The last category of "fictitious accounts" included the "Stock," "Profit and Loss" and "Factory" accounts. These accounts were imperative to an understanding of the economic position of the firm even though no single tangible document or item of value supported the account.[21] An important point to keep in mind is that each transaction involved at least two of these various accounts. The accounts could be from the same category or from different categories, depending on the particular transactions involved.

Each account was divided into two sides. To debit an account meant simply to place the information on the left

22

hand side of the account. To credit the same account was to place the information on the right hand side. Whether debiting or crediting an account increased or decreased it depended on the particular account involved. For the "personal accounts" and "real accounts" each transaction was analyzed in terms of the recognition of something of value in order to determine the debit(s) and credit(s). Thus, for example, an account such as a receivable, inventory or cash account was debited to increase the account, and credited to decrease the same account. For an account such as a payable, capital or factorage account, however, the opposite occurred. Values were derived from the purchase or sale of goods or services and the recognition of promises to pay or deliver by the company, or by some other party. The "fictitious accounts," on the other hand, were credited for increases in the account, and debited for decreases. Values were derived from the recognition of revenues, expenses, profits and losses. In all the entries, the accounts were "Debtor(s) to" or "Creditor(s) by" some other account.

When Duplanty introduced double-entry bookkeeping techniques into the DuPont Company record books, he accomplished this metamorphosis only after spending considerable time reviewing, correcting and augmenting Bauduy's efforts. He incorporated the information Bauduy had documented into his more sophisticated system with some difficulty. He encountered difficulty because he had to annotate existing account books and create other essential volumes in order to institute the double-entry system. In the process, he established the intricate interrelationships between the record books indicative of the double-entry bookkeeping system.[22] From the record keeping perspective, this was the event of most significant benefit to the firm. Duplanty finally made it possible to understand the firm's economic position.

A general description of the interrelationships between account books for a particular firm must, of necessity, identify types or categories of account books. Individual anomalies within the record keeping system, although significant to the firm in terms of content or impact, were not representative of the overall process. The following description was limited, therefore, to DuPont's generally recognized record books.

The Manuscripts Department of the Eleutherian Mills Historical Library (now Hagley Library) previously designated the specific account books which fell within the bookkeeping categories identified. The nomenclature assigned by the personnel in the Department depended, in part, on the references in other documents to the particular volumes in question. Some volumes were not cross referenced, however. The category Blotter included many such volumes. The Manuscripts Department therefore used the term almost as a generic label to encompass a miscellaneous assortment of account books. Strict interpretation of the classification limits the number of volumes that conform to the generally recognized definition of the function of a Blotter, and therefore the term did not apply to all the volumes arranged under that label.

The entries in the volumes that actually qualified as Blotters were rough, day-to-day notations of each and every event of the day as it happened.[23] This rather crudely kept account book represented the book of original entry for the transactions affecting the firm. The entries were not as carefully written as those later carried to either the Waste Book or the Journal, depending on which of these was maintained, but they sometimes contained far more detail. The information in the Blotter can be traced to the appropriate Waste Book, Journal and Ledger by correlating the entries with those in the other volumes. Since few, if any, references were made to the Blotters in other account books some of the volumes may no longer exist. This was relatively unimportant, however, as the information in any of the missing volumes was contained in at least one of the other account books and was often duplicated by the entries in the associated Waste Book and Journal. The Manuscripts Department of the Hagley Library catalogued seven volumes as Blotters for this period but only four of these qualified under that label.[24]

Contemporary authors recognized the Waste Book as the first of the primary account books to be maintained by a firm.[25] It was the book of original entry if there was no Blotter for the period. If a Blotter existed, the information it contained was transferred or posted to the appropriate Waste Book. The information in the Waste Book was then posted to the appropriate Journal. The entries were not always exactly duplicated because at times efforts were made to reorganize the material as it was posted from the Waste Book to the Journal. The Waste Book was maintained in a general Journal entry format. The information was entered following a heading designating the account names involved in the event.[26] The Waste Book entries were not annotated in any definite way to indicate posting to the appropriate Journal or Ledger. Sporadic marks on or next to the entries might have been efforts to document or verify the posting process, however. There were two existing Waste Books which covered the period under discussion.[27]

The Journals duplicated the information in the Waste Books except that the entries were annotated with the folio or page number of the specific account locations in the Ledger.[28] Each Journal entry was complete unto itself, and contained all the information necessary to post the entry to the Ledger. The entries in the volumes were, for the most part, chronologically ordered. Lapses from this format late in the period appeared to be efforts to reorganize the chronological material from the appropriate Waste Book.[29] For the period under consideration, there were four Journals.[30]

The Ledgers were large, heavy volumes to which all the information amassed in the various supporting volumes was ultimately posted.[31] Each Ledger was organized by account name and associated folio number. The function of this category was to present, in an organized format, all the information applicable to each account within the firm which was posted from the Waste Books or Journals.[32] The entries in the Ledger might be annotated with both the Journal folio number equating to the page location of the information which had been posted to the Ledger and the folio number equating to the location of the offsetting account(s) in the Ledger. The Ledgers were prepared in a standard format whereby either the left page was dedicated to the Debit entries and the right page to the Credit entries, or one page was divided in half lengthwise with both Debit and Credit entries on the same page. In addition, when either side of an account was filled up in one part of the Ledger, other empty or partially empty pages might be used for the additional information. The indices traced this path, and there were also forwarding notations on the full pages. There were three Ledgers for this period.[33]

Three general categories of secondary account books were also important in understanding the record keeping in the DuPont Company at this time. There was much less agreement in contemporary literature as to the kinds of supporting or secondary records to be maintained by a firm.

The books in the library archives under the overall title Accounts Current served a significant support purpose with respect to the firm's account books. These books were used to maintain a running account for certain key and active agents in the field and also, apparently, for individuals or firms with very active accounts.[34] At times these volumes were also used to store additional information, as something of a Memorandum volume.[35] The information in the earliest volumes could not be correlated easily with specific notations in other account books. Later on, however, transactions involving specific accounts included in the Accounts Current record books could be traced to other account books. There were four books within this category which covered this period of time.[36]

The Cash Books were used to document the cash disbursements of or for the company. The record keepers did not record any cash receipts in these volumes. Such receipts were recorded directly into the book of original entry. The bookkeepers recorded the checks that were written on behalf of the DuPont Company as well as the few direct cash payments. The information that was recorded in the Cash Books identified the recipient and, sometimes, the purpose of the check. Columnar headings attributed the expenditures to various firm components as applicable.[37] This information was then posted to the Journal, most likely at the end of the month. Initially, the only bank with which the DuPont Company had an account was the Bank of Delaware. Toward the end of this period, however, there were three banks with which the firm did business: The Wilmington & Brandywine Bank, The Farmers Bank, and The United States Bank. There were two books within this category that applied to this period.[38]

The Petit Ledgers were used by company record keepers to record the employees' accounts with the firm. These accounts included wages owed to the workers, when and how the wages had been earned, and any payments made either in cash or some good or service for or on behalf of the individual. These volumes were not considered a primary component of this research, but they were systematically maintained account books within the DuPont Company's early

27

record keeping system. Four volumes qualified as Petit Ledgers during this period.[39]

In modern day society, Certified Public Accountants verify the financial statements and underlying transactions of publicly-traded corporations. Such supervision and verification was not, of course, available to early nineteenth-century firms. Nonetheless, the record books of the early DuPont Company were essential to an understanding of the operations, worth and progress of the enterprise. Unfortunately, however, in a period when no overseeing authority supervised either the methods used in keeping the books, or the backgrounds of the bookkeepers, some of the information contained in these volumes was contradicted by other sources. This was particularly true in the case of the timing, nature, and ultimate value of the capital investment in the gunpowder manufactory during the period under consideration.

Notes--Chapter One

1   Numerous publications on bookkeeping were published
     prior to 1800. See Institute of Chartered Accountants
     in England and Wales. Historical Accounting Literature
     (London: Mansell Information/Publishing Ltd., 1975).

2   See the "Annotated Bibliography" for the
     identification, description and specific discussion of
     the general account books considered pertinent to this
     research. Examples from these volumes were altered to
     conform to current spelling, punctuation and
     capitalization norms for purposes of clarity. The
     substance of the original was maintained at all times.

3   Peter Bauduy was the first record keeper. Raphael
     Duplanty succeeded him as primary record keeper,
     although Bauduy continued to participate in the
     bookkeeping. The handwriting of both of these
     individuals was distinctive, and readily identifiable.
     There were other unidentified record keepers as well.

4   See appendix A.

5   Individuals involved in management training at the
     Administration of Powder and Saltpetre held a number
     of different managerial positions. There is no
     indication that trainees were ever directly
     responsible for the record keeping, however, other
     than for the personal records of their stipends and
     expenses. L'Assemblée Nationale. "Loi Relative à la
     Fabrication & Vente des Poudres & Salpêtres." Paris:
     De l'Imprimerie Royale, 1791. In addition, a closer
     examination of the translation by B. G. DuPont of a
     letter from Pierre to E. I. suggests that an
     alternative translation should be considered.
     DuPont's translation reads "I would . . . rather have
     you employed at book-keeping at the Arsenal." P. S.
     DuPont to E. I. DuPont, April 17, 1790. DuPont, Life,
     I:122. Instead, with the management training in mind,
     the meaning of the letter should be interpreted as "I
     would . . . rather have you employed in a position of
     responsibility at the Arsenal." Longwood Manuscripts,
     Group 1, #51. See Nicolas Gouin Dufief, A New
     Universal and Pronouncing Dictionary of the French and
     English Languages (Philadelphia: T & G. Palmer, 1810)
     and John Garner, Le Nouveau Dictionnaire Universel.

Vol. 1: <u>French--English</u> (Rouen, France: Pierre Dumesnil et Fils, 1802) for the translation. The idea of administration does not preclude some familiarity with bookkeeping, of course, but it does support my contention that E. I. was not a practiced bookkeeper.

6   E. I. maintained the record of powder sales, volume #1640 and its successor volume, #1643. Accession 500, Powder Sales Book, #1640. Accession 500, Powder Sales Book, #1643. Not in Annotated Bibliography. There is evidence that he also, occasionally, reviewed the books and annotated or generated some of the entries. He did not perform a complete audit of the books, however.

7   Dorothy Garesche Holland, <u>The Garesche, de Bauduy, and des Chapelles Families</u> (Saint Louis: Schneider Printing Company, 1963), p. 19-24 (hereafter cited as Holland, <u>Families</u>).

8   The articles of agreement between the major shareholder, DuPont de Nemours Father, Sons & Company, and Peter Bauduy were dated August 25, 1802. Untitled. DuPont, <u>Life</u>, VI:108-9.

9   Prior to this he was either out of town or negotiating his agreement with E. I. DuPont. See DuPont, <u>Life</u>, VI:74-80 for the negotiations between E. I. and Bauduy. For indications of Bauduy's absence, see E. I. DuPont to Peter Bauduy, June 25, 1802 and E. I. DuPont to William Hamon, July 10, 1802. DuPont, <u>Life</u>, VI:73-4 and 80-82.

10 Bauduy generated all the entries in the firm's record books until February 11, 1806, when Duplanty's influence became apparent. Accession 500, Blotter, #848, folio 51. Accession 500, Journal, #877, folio 74.

11 He continued to maintain the primary account book after February 11, 1806, for instance. Duplanty did not maintain the Journal until after December 15, 1808. Accession 500, Journal, #877, folio 200.

12 Riggs, <u>Guide</u>, p. 75. Accession 1360, p. 45.

13 Riggs, <u>Guide</u>, p. 75. Duplanty apparently generated a record keeping sampler using dates in 1802 sometime prior to his official affiliation with the firm. This probably represented an attempt to assist Bauduy in keeping the accounts. Winterthur Manuscripts, Group 4, Series D. His handwriting does not appear in the

actual record books with any regularity until after February 11, 1806, however.

14   Accession 500, Blotter, #848, folio 52.   Accession 500, Journal, #877, folio 74.

15   The principal account books were the Waste Book, the Journal, and the Ledger.

16   Accession 500, Journal, #877, first page, unnumbered. (For a complete discussion of the timing of shareholder investments, see Chapter Two.)

17   This process was not introduced until February 11, 1806 when the information in the Factory Building Book (#848) and the Factory Book (#850) to that date was carried to the Journal, Duplanty numbered the accounts throughout the Journal, and the Ledger was created. Accession 500, Blotter, #848, folio 52.   Accession 500, Blotter, #850, folio 15.   Accession 500, Journal, #877, folios 71 & 73.   Accession 500, Ledger, #934.

18   John B. Geijsbeek, trans. <u>Ancient Double-Entry Bookkeeping. Lucas Pacioli's Treatise (A.D. 1494-the Earliest Known Writer on Bookkeeping) Reproduced and Translated</u>. (Denver, Colorado: J. B. Geijsbeek, 1914).

19   Mair, <u>Moderniz'd</u>.   Dilworth, <u>Assistant</u>.

20   These accounts were annotated with the terms "Proper," "in Company," or "Factorage."   The Account Proper meant an active account with a firm.   The Account in Company was the investment account.   The Factorage Account was an agent's account with a firm.   Dilworth, <u>Assistant</u>, Annex, first page, unnumbered.   Accounts could also be labelled "Account Current."   Proper and Current accounts were essentially the same.

21   Mair, <u>Moderniz'd</u>, p. 20.

22   Accession 500, Journal, #877, folios 71 & 73. Accession 500, Blotter, #848, folio 52.   Accession 500, Blotter, #850, folio 15.   The Ledger was maintained by Duplanty and therefore would not have been generated until he assumed primary responsibility for the record keeping as of February 11, 1806. Accession 500, Ledger, #934.

23   The <u>Oxford English Dictionary</u> definition of a Blotter stated that it was "a term applied in counting houses to a wastebook."   The reference was dated 1847 and there were no earlier references to this term in the

31

dictionary or the supplement. James A. H. Murray, ed.
<u>A New Oxford Dictionary on Historical Principles</u>
(Oxford: Clarendon Press, 1888). R. W. Burchfield,
ed. <u>A Supplement to the Oxford English Dictionary</u>
(Oxford: Clarendon Oxford England Press, 1972).
Certainly, the contemporary editions of the popular
works cited here did not mention the term "Blotter."
Mair, <u>Moderniz'd</u>. Dilworth, <u>Assistant</u>. This did not
mean, however, that it was not in use at this time as
an independent, rough account book of original entry.

24 Accession 500, Blotter, #851, 852, 853 and 854 were
actually blotters. The functions of all seven volumes
are discussed in detail in the Annotated Bibliography.

25 Mair, <u>Moderniz'd</u>, p.4-5.   Dilworth, <u>Assistant</u>,
Preface--unnumbered, p.5.

26 Individual account names were used in the heading, as
well as "Sundries Dr. to Cash," for example, or "U. S.
Bank Dr. to Sundries." The accounts included in the
term Sundries would then be detailed in the entry
itself.

27 Accession 500, Waste Book, #862, 863. There are
references to another, earlier Waste Book on July 20,
1809, for example, in the Journal for that period
which indicated "as per minute a/c in waste book," but
this volume apparently no longer exists. Accession
500, Journal, #878, folio 13.

28 Mair, <u>Moderniz'd</u>, p. 8. Mair saw the Journal as "the
book wherein the transactions recorded in the
Waste-book are prepared to be carried to the Ledger,
by having their proper debtors and creditors
ascertained."

29 The bookkeeper's attempts to organize the Journal
meant that the transactions were not recorded as
complete entries with both debits and credits.
Rather, the information for each account name was
disaggregated and placed in separate debit and credit
groups dated at the end of the month. Over time, the
rigor of this practice decreased. See Accession 500,
Journal, #880, folio 138-234.

30 Accession 500, Journal, #877, 878, 879, 880.

31 Mair described the Ledger as "the Waste-Book taken to
pieces, and put together in another order: the
transactions contained in both are the same, but

recorded in a different manner." Mair, Moderniz'd, p. 3.

32 Mair said of the Ledger that "things of the same kind are classed together, and all the particular items and articles belonging to the same subject are collected and united." Mair, Moderniz'd, p. 3.

33 Accession 500, Ledger, #934, 935, 936.

34 There was also information on interest owed to shareholders in the firm, such as Necker-Germany, Duquesnoy and Bidermann. Accession 500, Accounts Current, #1064, folios 27 & 36. There were also references in the various Accounts Current volumes to certain firms with which the principals in the DuPont Company had close ties. Therefore, such firms as Bauduy, Garesches and Company, and DuPont, Bauduy and Company were included in these volumes. Accession 500, Accounts Current, #1064, folios 58 & 67.

35 Information on Treasury notes received as payment was included in this volume, for example. Accession 500, Accounts Current, #1064, folio 59.

36 Accession 500, Accounts Current, #1064. Other volumes that apply to the period under consideration had no bearing on this research and were not included in the Annotated Bibliography.

37 E. I. DuPont, Peter Bauduy, and other firm components. Accession 500, Cash Book, #1035.

38 Accession 500, Cash Book, #1035. Other volumes that applied to the period under consideration had no bearing on this research and were not included in the Annotated Bibliography. There was also reference to a Cash Book which is no longer extant.

39 The designated Petit Ledgers were not included in the Annotated Bibliography, but they resembled the description of payroll records in Accession 500, Blotter, #849, in Annotated Bibliography.

Chapter Two

THE CAPITALIZATION

The bookkeeping entries made by Bauduy to record the sequence of events leading to the initial capitalization of the DuPont Company were generally inaccurate and misleading. Early investors in the DuPont Company could not provide the firm much direct cash investment. Instead, they offered alternative financing arrangements either by taking care of obligations owed by the firm, or undertaking some other means of completing their investment promises. Bauduy did not detail all of these transactions, however. Nonetheless Duplanty incorporated this sequence into the Ledger he created in February 1806. Duplanty was more successful in his documentation of subsequent events and their aftermath which caused dramatic changes in the capitalization of the gunpowder manufactory, however. Bauduy's disenchantment as a partner and eventual withdrawal from the firm and the addition of numerous petty shareholders due to the dissolution of DuPont de Nemours Father, Sons & Company were significant events well documented by the firm's record keepers.

When E. I. convinced his father that the operation of a gunpowder manufactory might solve the family's fiscal concerns, he had only taken the first step on the long road to success. From the beginning the DuPonts considered the prospective gunpowder manufactory essentially an offshoot, or branch, of Pierre's firm, the "parent company," DuPont de Nemours Father, Sons & Company.[1] The family wanted the parent company to remain in control of all operations and decision making for the new firm.[2] Therefore E. I. had to deal with several problems associated with the need to fund the new enterprise almost immediately.

E. I. had already decided to capitalize the gunpowder manufactory for $36,000, and had divided this amount into eighteen shares of $2,000 each.[3] He believed this amount would adequately finance the start up and early operating costs of the enterprise. The family determined that two-thirds of the invested capital, or twelve of the shares, would have to be funded by the parent company for control of the gunpowder manufactory to remain in the hands of the DuPonts.[4] Unfortunately, the parent company could not provide enough investment capital out of available funds to purchase the necessary shares in the firm.[5] The family therefore devised a scheme whereby an investor in one share of the new firm also purchased two additional shares of the same value in the parent company. Pierre would then use this additional investment to purchase two shares in the gunpowder manufactory in the name of his firm. If the plan worked, a six-share investment in the gunpowder manufactory would automatically fund the parent company's planned investment in the firm.[6] This stipulation probably increased the difficulties E. I. faced in the process of funding his endeavor, largely because his father's firm was not successful. In addition to this, however, other circumstances also served to complicate his quest for funding.

When E. I. did finally concern himself with the funding for his new enterprise he turned to the investment sources he and his family knew best, the French. The French investors he approached included individuals who had already invested in his father's firm, as well as those who had only promised support or expressed interest in that venture.[7] E. I. had little to present to these potential investors, however, other than his plans for the gunpowder manufactory.[8] He therefore urged potential backers to take

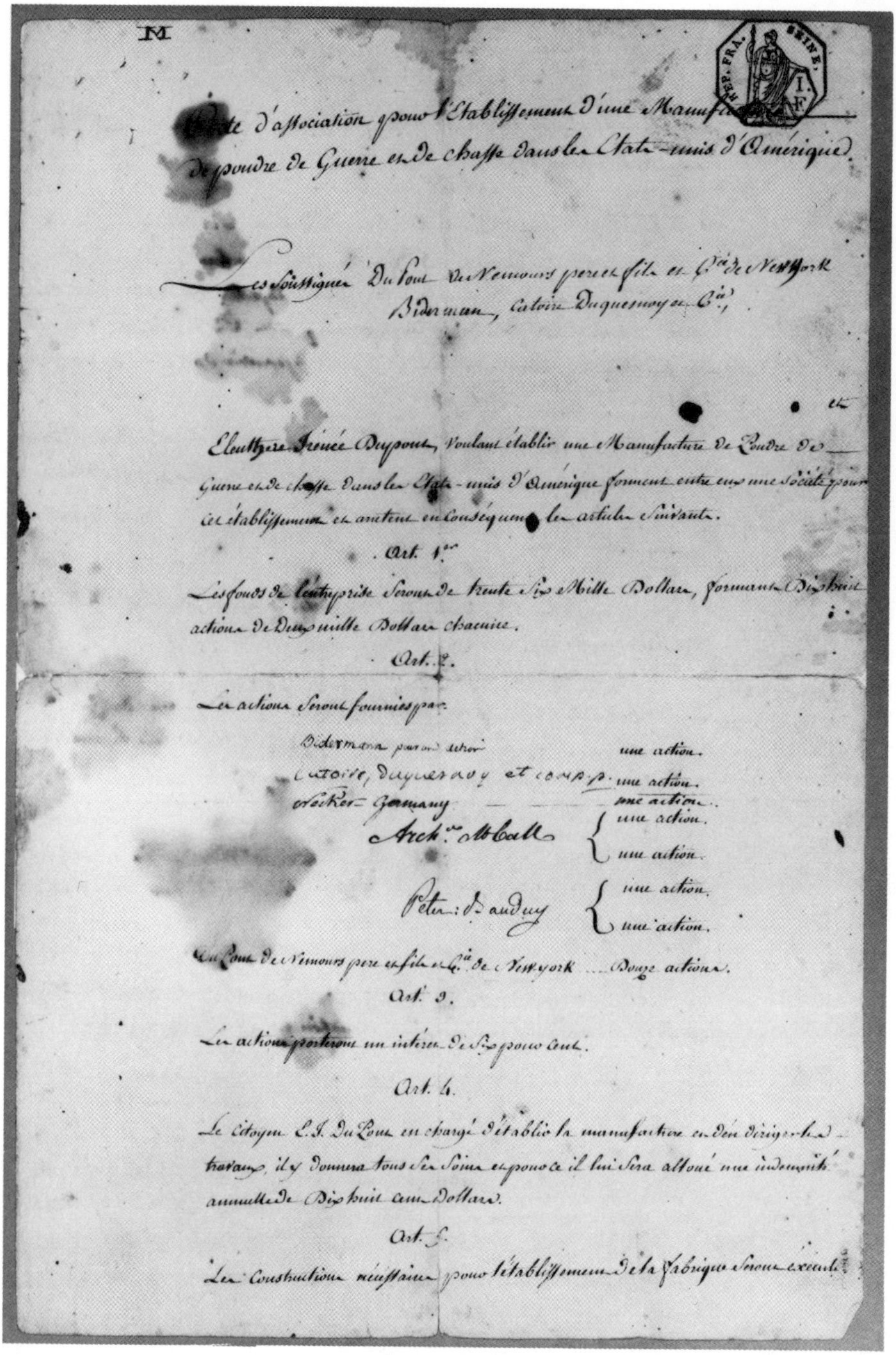

Illustration 4: An excerpt from the original Deed of Association showing the shareholders' signatures. Accession 146, File 21. Photograph courtesy of Hagley Museum and Library.

a chance on an unseen, untried enterprise located overseas and virtually inaccessible to the French investors.[9] The fact that E. I. had little success meeting the capitalization requirements he envisioned for the firm belied documented evidence to the contrary, however.[10]

The Deed of Association, for instance, listed several shareholders in E. I. DuPont de Nemours & Company as of April 21, 1801.[11] The fact that this document existed, however, did not mean that the shareholders' signatures had been obtained on that date.[12] The document was in all likelihood prepared on April 21, 1801, and yet remained unsigned by the shareholders for some time.[13] The first account book of original entry also provided conflicting information concerning the initial capitalization of the firm.[14] The entries not only contradicted the evidence presented in the Deed of Association, but also included some additional misleading information.

The actual investment sequence differed considerably from the documented sequence implied in both the Deed of Association and the first account book.[15] Indeed, E. I. returned from France in July 1801 with little success to report in his efforts to fund the gunpowder manufactory.[16] At that time he had only attracted two investors for the firm: Jacques Bidermann and the firm, Catoire, Duquesnoy & Company. In fact, these were the only original investors who were in a position to sign the Deed of Association on April 21, 1801.[17]

Jacques Bidermann, a successful Parisian banker born in Switzerland, had been the largest investor in Pierre's original firm.[18] He was also in all likelihood the first person E. I. contacted when he arrived in France in early 1801. Even though Bidermann was well aware of the suggested changes in the objectives of Pierre's firm when E. I. approached him concerning the funding for the gunpowder manufactory, he agreed to support the

37

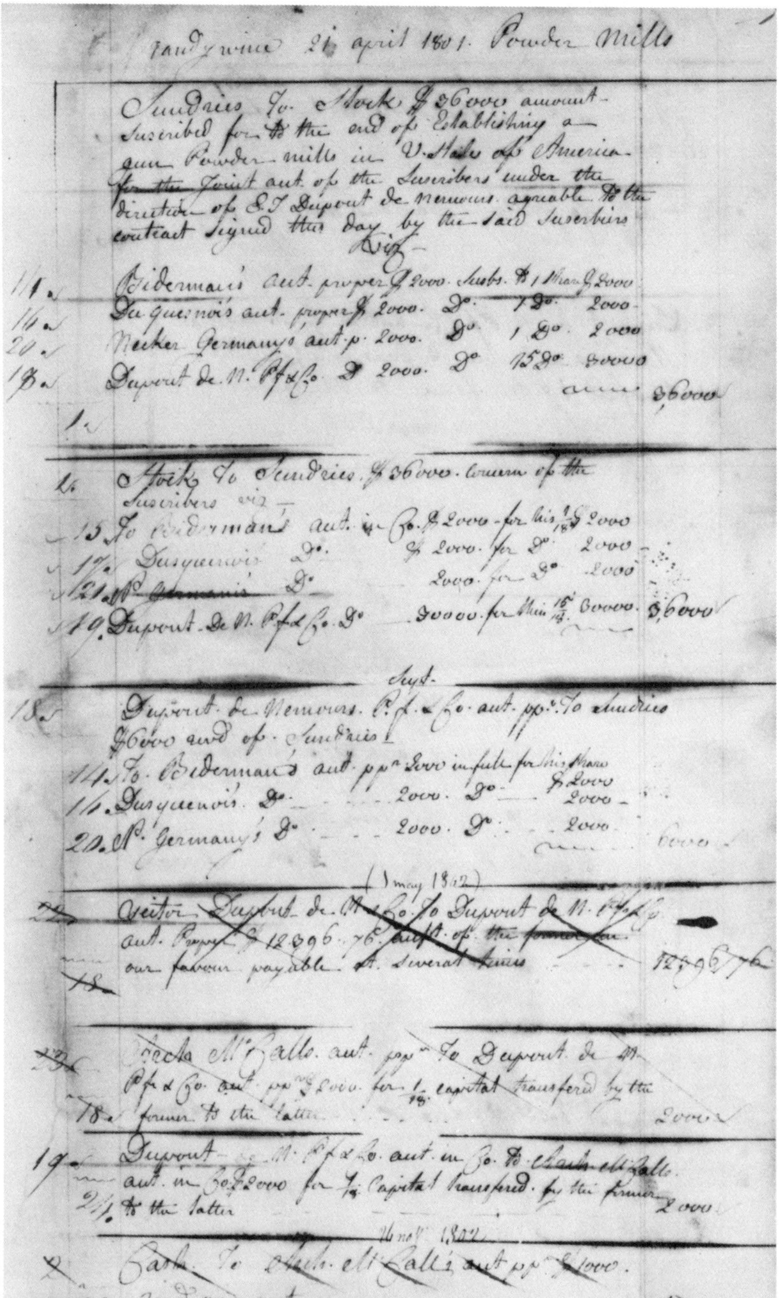

Brandywine 21 april 1801. Powder Mills

Sundries To Stock $36000 amount subscribed for the end of Establishing a gun Powder mill in U. States of America for the Joint acct. of the Subscribers under the direction of E. I. Dupont de Nemours agreeable to the contract signed this day by the said Subscribers Viz—

Bidermann's acct. proper $2000 Subs. to 1 May $2000
Du Quesnoi's acct. proper $2000 Do 1 Do 2000
Necker Germany's acct. p. 2000 Do 1 Do 2000
Dupont de N. P. & Co. Do 2000 Do 15 Do 30000
36000

Stock To Sundries $36000 amount of the Subscribers viz—
15 To Bidermann's acct. in Co. $2000 for his 1/18 $2000
17 Dusquenoi Do $2000 for Do 2000
21 Mr. Germany Do 2000 for Do 2000
19 Dupont de N. P. & Co. Do 30000 for their 15/18 30000 36000

Dupont de Nemours P. & Co. acct. ppr. To Sundries $6000 recd of Sundries
14 To Bidermann's acct. ppr. 2000 in full for his share $2000
14 Dusquenoi Do 2000 Do 2000
20 Mr. Germany's Do 2000 Do 2000 6000

(1 may 1802)
Victor Dupont de N. & Co. To Dupont de N. P. & Co. acct. Proper $12896.76 acct. of the former in our favour payable at several times 12896.76

Jacob McCall acct. ppr. To Dupont de N. P. & Co. acct. ppr. $2000 for 1/18 capital transferd by the former to the latter 2000

Dupont de N. P. & Co. acct. in Co. To Charles McCall acct. in Co. $2000 for 1/18 Capital transferd by the former to the latter 2000

26 nov. 1802
Cash To Arch. McCall's acct. ppr. $1000
recd on acct 1000

Illustration 5: The first page, unnumbered, in the first record book. The first two entries constitute the capital entries. Accession 500, Journal, #877. Photograph courtesy of Hagley Museum and Library.

enterprise.[19]   He agreed to purchase one share in the gunpowder manufactory directly, and fund two shares in the name of the parent company.[20]   Bidermann did not invest cash directly in the firm.   Rather, the proceeds of his investment were used to pay for materials and machinery E. I. had bought in France, and any remainder was to be remitted to the company.   It was not clear when and if the shares were ever funded, although apparently payments were completed.[21]   Bidermann's signature appeared on the Deed of Association and his name was included in the first record book.[22]   He continued his affiliation with the firm throughout the period under consideration.

The firm Catoire, Duquesnoy & Company probably agreed to invest in the gunpowder manufactory at least in part because of the long standing friendship between Pierre and Adrien Cyprien Duquesnoy.[23]   The firm agreed to the investment stipulations presented by the family while E. I. was still in France.[24]   Again, this investment did not directly involve cash.   Instead, the firm apparently planned to fund its investment with the proceeds from the sale of a load of salt.[25]   As with Bidermann, it was not clear when and if the shares were ever funded.[26]   The name of the firm appeared on the Deed of Association and Duquesnoy's name alone appeared in the first record book entry.[27]   Duquesnoy remained a shareholder in the firm until December 31, 1809.[28]

Louis Necker de Germany (hereafter Necker-Germany) was also recognized on the original Deed of Association and in the first record book entry as the owner of one share of stock in the gunpowder manufactory.[29]   Again, the timing of his investment, and of his signature, was not clear. Necker-Germany had only expressed an interest in the enterprise and had not purchased any shares in the gunpowder manufactory while E. I. was still in Paris. E. I. continued to hope that he would invest in the firm in accordance with the terms for such investment, however.[30] Unfortunately Necker-Germany in his turn attached his own stipulations to the investment.[31] He did finally agree to take one share in the gunpowder manufactory. Again, it was not clear when and if the shares were ever funded.[32] Necker-Germany continued his affiliation with the firm until his death, when the responsibility for the realization of his investment fell to his survivors.[33] The account for Necker-Germany therefore remained open until May 1808 when this share was transferred to E. I.[34]

The only other significant early investment in the gunpowder manufactory was an unexpected windfall that greatly benefited the young firm. On August 8, 1801, Victor indicated that Jacques Necker, the former French Finance Minister, had offered to loan the family $9833.40 to be used for the gunpowder manufactory.[35] The proceeds of this windfall were not treated as a direct investment in the gunpowder manufactory, however. Rather, the loan was treated as an additional investment by the parent company in five shares of the gunpowder manufactory.[36] The repayment schedule for the loan was not clear.[37] Necker died on April 9, 1804, in Coppet, Switzerland. His daughter, Anne Louise Germaine Necker, Baronne de Staël-Holstein (Mme de Staël) inherited the right to collect the loan.[38] This loan, although so badly needed at the time, would eventually cause E. I. and his father a great deal of trouble.[39]

As is evident, E. I. preferred French investors for his enterprise, and was leery of Americans. He nonetheless eventually had to accept an American offer of support from Archibald McCall, a merchant in Philadelphia. On June 1, 1802, one-eighteenth of the capital ($2000) credited to the parent company was transferred to McCall.[40] Apparently at the same time his name was recorded on the original Deed of Association and he was accorded two shares in the firm on

40

that document.[41]   This discrepancy probably reflected the problems created by McCall's interest in the company.   He actually preferred a more sizable investment in the firm and wanted at least three shares.[42]   His offer was not immediately accepted because the DuPont brothers recognized the benefits of minimizing the number of different shareholders in the firm and also realized that no investors "outside" of the family should achieve the upper hand in any negotiated transaction.   McCall's preference continued to occupy the brothers, however.[43]   McCall paid $1000 to the company on November 26, 1802, at least five months after he expressed his intention to invest in the firm.[44]   He remained a shareholder until September 5, 1803, when his share of stock was transferred to Peter Bauduy and the $1000 was returned to him.[45]   Nonetheless, McCall continued to function as an agent, contact, supplier and creditor for the gunpowder manufactory despite this sequence of events.[46]

Another early investor in the gunpowder manufactory, William Hamon, became affiliated with the firm in 1802.   At that time, E. I. decided to purchase land along the Brandywine River in Delaware as a location for his firm. He was concerned, however, that restrictions existed against foreigners buying land in their own names.[47]   E. I. therefore decided to find an American citizen to make this important purchase.   The person he chose, William Hamon, was a naturalized citizen from Santo Domingo and the brother-in-law of Peter Bauduy.[48]

Hamon also agreed to purchase one of the firm's shares in June 1803.   In all, he paid $1500 toward his subscription in June and August 1803.[49]   He was forced to give up his interest in the manufactory in September 1803.[50]   His share of stock was transferred to Peter Bauduy on September 9, 1803.[51]   Hamon's account with the company continued for some time, however, and despite his financial reverses he was apparently in no rush to recoup his investment in the firm.[52] His name did not appear on the Deed of Association.[53]

Peter Bauduy, DuPont's first record keeper, was also, in the early years, the most important investor in the gunpowder manufactory. This distinction applied because his investment was not restricted to the purchase of shares, but included the extension and arrangement of credit opportunities imperative to the continuance of the firm. In one way and another he had a decided impact on the health and future both of the founder of E. I. DuPont de Nemours & Company, and of the enterprise itself.

Bauduy first became involved with the firm when E. I. was looking for a location for his mills.[54] Once E. I. had chosen Wilmington as the optimal location for the manufactory, he and Bauduy worked closely together to locate the gunpowder works along the Brandywine River.[55] Having helped the DuPont family locate in Wilmington, Bauduy also wanted to participate in the enterprise. The correspondence between Bauduy and E. I. in mid-1802 indicated that E. I. believed Bauduy's conditions unreasonable.[56] He did not rule out the possibility of a partnership, however, especially since Bauduy offered him so much support in funding the firm.[57] Victor strongly encouraged an association with Bauduy as well.[58] After much negotiation, Bauduy became a partner in the firm on August 25, 1802. His terms were delineated in separate Articles of Agreement drawn up with the parent firm. In exchange for his investment Bauduy received two shares in the enterprise, three of the eighteen shares of the profits earned by the firm, and a two and one-half percent commission on powder sales.[59] This investment was also recorded on the original Deed of Association.[60] In addition, Bauduy managed to purchase the shares that McCall and Hamon were selling.[61] Once again, none of these transactions involved cash.[62]   Despite Bauduy's value to the firm, from the early days the relationship was strained.

E. I. had gone into the manufactory not as an owner, but as a manager. His past experience as a management trainee at the Administration and as manager of his father's print shop in Paris taught him management skills that particularly qualified him to fill this position. He therefore was designated the director of the manufactory in the Deed of Association.[63]   The director's responsibility was to establish and superintend the day-to-day operations of the business to the best of his ability for the benefit of the shareholders, and particularly for the benefit of the majority shareholder in the gunpowder manufactory, considered from the very beginning of the enterprise to be DuPont de Nemours Father, Sons & Company.   E. I.'s interpretation of his responsibility to the shareholders was probably based, in a large measure, on his understanding of the organizational construct of the firm.

His father's firm had been organized as a "societé en commandité." A "societé en commandité" involved several "sleeping partners" who invested money in the venture, but had no managerial rights or responsibilities.   Such an organization could guarantee limited liability to these silent investors.[64]   The family believed that the same business forms would apply in the United States as well.[65]

Such was not the case, however.   In the United States, the limited partnership did not exist at this time.   All recognized partners in a firm were subject to unlimited liability.[66] Bauduy understood this, and therefore held an entirely different view of his responsibilities in the organization. Whereas E. I. considered himself entirely in charge, Bauduy as a partner realized his potential liability for the obligations of the firm, and therefore wanted a more active role in operations.   These opposing points of view became apparent early in the firm's history and caused a great deal of ill feeling and discord which pervaded all the years of the partnership between E. I. and Bauduy.

The first evidence of trouble between the two concerned a serious disagreement in October 1804 involving an advertisement placed by McCall in a Philadelphia newspaper. The incident erupted into a conflict which almost destroyed the new firm.[67] The conflict was finally resolved, at least in part, because E. I. drew up a new agreement with Bauduy dated July 1, 1805.[68]

In the Deed of Association the termination date for the first partnership was January 1, 1810. The original shareholders were to indicate their intentions with respect to the firm--whether to stay owners or not--before January 1, 1809. Arrangements were made to allow shareholders to withdraw without dissolution of the partnership at that point if less than one-third of the firm's owners chose that option. The disenchanted shareholders would receive a return of their investment with an additional 6% interest for the duration of their ownership.[69]

On December 31, 1809, bookkeepers recorded the shareholders' investment interest immediately prior to the new agreement. DuPont de Nemours Father, Sons & Company was accorded interest on eleven shares, Bidermann interest on one share, Duquesnoy interest on one share, Necker-Germany interest on the share he owned through May 1, 1808 when it was purchased by E. I., and E. I. subsequent interest on Necker-Germany's share.[70] Bauduy received interest on the four shares he owned at this time although it was not specifically identified.[71] Catoire, Duquesnoy & Company sold their share in the firm to the parent company as of this date.[72]

The new partnership began on January 1, 1810 and was scheduled to last for nine years.[73] On that date the

44

record keepers recorded the following holders of "Shares in the primitive Stock" in the Waste Book: Bidermann (one share), E. I. DuPont (one share), Peter Bauduy (four shares), and Dupont de Nemours Father, Sons & Company (twelve shares).[74] Several changes in the ownership of the company occurred in the next few years, however.

The parent firm to the gunpowder manufactory, DuPont de Nemours Father, Sons & Company, was originally scheduled to terminate in July 1811 or 1812.[75] The company was actually dissolved on May 20, 1811.[76] The only assets owned by the firm of any value were the shares in the gunpowder manufactory. As early as April 1808 Pierre had allocated the shares his firm owned in E. I. DuPont de Nemours & Company among his shareholders, and effectively identified that these shareholders would be given the option of receiving the principal plus interest on their investment, or of becoming shareholders in the gunpowder manufactory.[77] There were no indications on May 20, 1811, in the record books of the gunpowder manufactory, of the dissolution of the parent company or of the decisions made by the shareholders, however. Nonetheless, E. I. became responsible to his father's shareholders either as owners of the gunpowder manufactory or for the amount of their investment if they did not want to remain owners.[78]

During this period the relationship between E. I. and his principal partner did not go smoothly.[79] Despite the controversies and uneasiness between the two men, their families were close. Thus to further complicate matters Bauduy's son, Ferdinand, married E. I.'s daughter, Victorine, in late 1813. Tragically, Ferdinand died within weeks of the wedding in January 1814.[80] Apparently the alienating dissatisfaction experienced by Bauduy had only been temporarily banked over the intervening years, and had been germinating just below the surface. Perhaps because

45

of his son's death and the concomitant grief he experienced, in June 1814 the anger that Bauduy had repressed was released.[81] There was so much bitterness and controversy on both sides that Bauduy finally withdrew from the firm on February 28, 1815.[82]

Immediately prior to the documentation of Bauduy's withdrawal from the gunpowder manufactory, but still dated February 28, 1815, the record books of E. I. DuPont de Nemours & Company recorded the allocation of the twelve shares in the gunpowder manufactory to the shareholders of DuPont de Nemours Father, Sons & Company who wanted to retain their ownership interests.[83]

The events of February 1815 would occupy the DuPonts for the rest of the period under consideration, and for several years thereafter. There was much interaction between all of the partners in the gunpowder manufactory as they attempted to resolve conflicts within the firm. Bauduy was especially disturbed by the events and their aftermath. Eventually, he initiated a Chancery Suit which was dated April 18, 1817.[84] The account books were introduced as evidence in this suit which was not resolved until well after the period under discussion.[85]

The DuPonts never increased the gunpowder manufactory's capital base during the period in question. The family could therefore not anticipate possible cash inflows from outside investors. Instead, any additional cash flows depended on the firm's successful operations. This constraint was less significant than it would be in the late twentieth century, however. Modern demands for cash to fulfill operational requirements and meet interest payments did not apply in the early nineteenth century. Cash flows were not nearly as important in the credit-based economy of that period.

Notes--Chapter Two

1   P. S. DuPont to Jacques Bidermann, December 1, 1801. DuPont, *Life*, V:192.

2   "Notes concerning the Powder Manufacture," undated. DuPont, *Life*, V:245.

3   This was recorded on the Deed of Association. See appendix A for the distribution and allocation of shares.

4   "Notes concerning the Powder Manufacture," undated. DuPont, *Life*, V:245.

5   DuPont de Nemours Father, Sons & Company, initially severely undercapitalized itself, had never been a successful venture and had no excess monies to invest. "Present Status of our Project," undated. DuPont, *Life*, V:99-109. P. S. DuPont to Jacques Bidermann, December 1, 1800. DuPont, *Life*, V:164-5, 194-5.

6   The family apparently devised this scheme before E. I. first attempted to find investors for his firm. Thus, when he did approach investors, he had to try to convince each one to invest at least $6,000 in three shares in the gunpowder manufactory, although only one of the shares would be allocated to that investor. "Notes concerning the Powder Manufacture," undated. DuPont, *Life*, V:245.

7   The same names show up over and over again either as actual or potential investors for the DuPont enterprises. "Present Status of our Project," undated. DuPont, *Life*, V:99-109. P. S. DuPont to Jacques Bidermann, December 1, 1800. DuPont, *Life*, V:163-196.

8   He prepared a general critique of gunpowder manufacturing in the United States and a description of the location he hoped to find and buildings he hoped to construct. "On the Manufacture of War and Sporting Powder in the United States," undated. DuPont, *Life*, V:198-205. "The Location and Constructions Necessary for Manufacture of Gunpowder," undated. DuPont, *Life*, V:206-212.

9   He probably also had to deal with another
complication. His brother Victor's responsibilities
at this time involved acquainting his father's
existing and potential investors with Pierre's
proposed revisions of the original objectives of his
firm. Unfortunately, some aspects of Pierre's
proposals for international commerce caused concern
among interested parties, including Victor himself.
Victor DuPont to DuPont de Nemours Father, Sons &
Company, August 8, 1801. DuPont, _Life_, V:254. The
fact that the gunpowder manufactory did not fall into
the same category as these other proposals probably
did not completely quell the perceived risk associated
with investment in a "DuPont family" undertaking.

10   For a complete understanding of the sequence of events
leading to the initial capitalization of the firm such
evidence must be considered in light of other,
relevant information. Detailed examination and
correlation of existing documentation concerning the
firm's capitalization highlighted the confusing morass
of information available on the early years of the
firm's existence.

11   See appendix A. The full text of the Deed of
Association as well as the share allocation were all
written in the distinctive handwriting of the
unidentified drafter of the document, probably a clerk
in a Parisian administrative department. The
shareholders' signatures were all apparently
authentic, except for those of Necker-Germany, and the
firm, DuPont de Nemours Father, Sons & Company. The
latter was also recorded by the clerk. Necker-
Germany's name and share allocation were obviously
inserted between two evenly spaced lines in E. I.'s
distinctive handwriting, however.

12   See appendix A. There is no clear indication on this
document of the dates of the original signatures which
were obtained at various times in both France and
America.

13   See appendix A. Not the least of the problems with
the signatures on the Deed of Association was that a
total of nineteen shares were allocated to
shareholders instead of the originally planned
eighteen. Certainly the twelve shares were
arbitrarily allocated to DuPont de Nemours Father,
Sons & Company prior to the time when they were
actually funded. In addition, discrepancies existed
between the actual timing and number of shares of the
shareholder investment and the subsequent recording of

48

the share ownership on the Deed of Association. Later additions to the manuscript clarified the situation somewhat, but did not completely explain the sequence of early events.

14   See appendix C for the first two entries in the first record book. They were recorded by Bauduy, and were dated as if entered on the day the firm was formed. Bauduy did not officially affiliate with the firm until August 25, 1802, however, and there is no reason to suspect that he had anything to do with the record keeping until after that time. Therefore, the capitalization entries dated April 21, 1801, in all likelihood reflected events as of the time Bauduy affiliated with the gunpowder manufactory, well after the date on the Deed of Association. Additionally, according to these first entries, the gunpowder manufactory as of April 21, 1801, was completely capitalized. This was not possible, however, since full funding for the firm had not even been identified at this point. "Notes concerning the Powder Manufacture," undated. DuPont, _Life_, V:245-6. Part of the confusion caused by these entries involved the allocation of fifteen shares in the gunpowder manufactory to DuPont de Nemours Father, Sons & Company as shown in the entries in appendix C. Certainly this allocation differed from the allocation on the Deed of Association. There are two possible interpretations of this allocation. One is that Bauduy, already described as an unsophisticated record keeper, simply allocated the shares in the only way he could to record the full capitalization of $36,000. It is also possible, however, that he recognized that the gunpowder manufactory as a branch of the parent company was considered "wholly owned," if not funded, by that company. Therefore, the assumption that all shares belonged to the parent company until purchased by someone else justified the allocation of any unsold shares to DuPont de Nemours Father, Sons & Company. For whatever reason, Bauduy used the right technique to document the events. Thus, at the time of an initial share purchase by new investors who were not included in the first record book entry, share values were transferred from the DuPont de Nemours Father, Sons & Company "Account in Company" to the new investor's "Account in Company." This practice applied to the investment by McCall, Hamon and Bauduy. The more sophisticated record keeper, Raphael Duplanty, carried the entries of such transfers to the first Ledger exactly as documented by Bauduy. Accession 500, Ledger, #934, folios 19, 24, 26, 28.

15   Therefore, to understand the firm's capitalization, one must consider the letters of this period in addition to the other documentation indicated. DuPont, *Life*, I-XII.

16   See the following letters for indications of his return. P. S. DuPont to E. I. DuPont, July 12, 1801. DuPont, *Life*, V:236. P. S. DuPont to Thomas Jefferson, July 23, 1801. DuPont, *Life*, V:238.

17   At this point Necker-Germany was the only other investor included on the Deed of Association who had expressed an interest in the firm. He and the other investors did not become actively involved until later. "Notes concerning the Powder Manufacture," undated. DuPont, *Life*, V:244-5.

18   "Present Status of our Project," undated. DuPont, *Life*, V:102.

19   Pierre sent E. I. and Victor to France with a letter addressed to Bidermann. In the letter Pierre suggested revising the original objectives of his firm, DuPont de Nemours Father, Sons & Company. P. S. DuPont to Jacques Bidermann, December 1, 1800. DuPont, *Life*, V:163-196.

20   "Notes concerning the Powder Manufacture," undated. DuPont, *Life*, V:244-5. Bidermann may have been impressed by the enthusiasm, knowledge and prior planning E. I. exhibited concerning the new firm. "On the Manufacture of War and Sporting Powder in the United States," undated. DuPont, *Life*, V:198-205. "The Location and Constructions Necessary for Manufacture of Gunpowder," undated. DuPont, *Life*, V:206-212. See also appendix B. He may also have seen the gunpowder manufactory as the only viable solution to the uncertain economic conditions facing DuPont de Nemours Father, Sons & Company.

21   "Notes concerning the Powder Manufacture," undated. DuPont, *Life*, V:250-1. According to the first record book, Bidermann paid DuPont de Nemours Father, Sons & Company "in full" for his one share investment in the gunpowder manufactory. Accession 500, Journal, #877, first page, unnumbered. The entry was dated September 1801, although this date was suspect since remittances on E. I.'s behalf in France were used, at least in part, to complete the transaction. Further, despite the entry in the first record book, completion of his promises took considerable time. E. I. DuPont to P. S. DuPont, June 12, 1802. DuPont, *Life*, VI:69. E. I.

DuPont to P. S. DuPont, August 15, 1802. DuPont, Life, VI:98. In addition, in some instances these payments were used by Pierre and therefore may never have reached E. I. E. I. DuPont to P. S. DuPont, August 15, 1802. DuPont, Life, VI:97. The timing and completion of the promised payments for the two shares to be purchased by DuPont de Nemours Father, Sons & Company was not clear since such payments did not show up in the record books attributed to Bidermann, and Pierre's records were woefully inadequate. See, for example, Longwood Manuscripts, Group 1, #470 and #478. The only evidence of completed payment is an undated memorandum. Longwood Manuscripts, Group 3, #2368. Bidermann was apparently a staunch supporter of the enterprise, so much so that when E. I. left France in 1801, he left the Deed of Association with him so that the signatures of additional investors could be added. "Notes concerning the Powder Manufacture," undated. DuPont, Life, V:244-5.

22 See appendices A and C.

23 Consequently, Duquesnoy was probably the individual in the firm E. I. approached for funding. This is also perhaps why Duquesnoy alone was recognized in the entry to record the investment. See appendix C.

24 "Notes concerning the Powder Manufacture," undated. DuPont, Life, V:245.

25 The plan, dated April 21, 1801 and signed by E. I. for DuPont de Nemours Father, Sons & Company, was for the salt to be shipped to the U. S. by the DuPonts and sold by them for a commission. The amount received from the sale was to be applied against three shares ($6,000) in the gunpowder manufactory if the return was sufficient to cover this amount. If the sale could not be effected by June 20, 1802, then the firm would place $4000 at the disposal of DuPont de Nemours Father, Sons & Company in France. When the salt was sold, the remaining share would be purchased and any excess remitted to the company. Winterthur Manuscripts, Group 2, #5416. The plan was not, apparently, signed by Duquesnoy or other representatives of his firm.

26 Duquesnoy's firm supposedly completed payments for at least one share in 1801. Victor DuPont to E. I. DuPont, August 13, 1801. DuPont, Life, V:266. Accession 500, Journal, #877, first page, unnumbered. The firm may also have agreed to finance two shares on behalf of Pierre's firm. The timing and completion of

the promised payments was not clear since such payments did not show up in the record books attributed to Duquesnoy or his firm, and Pierre's records were woefully inadequate. See, for example, Longwood Manuscripts, Group 1, #470 and #478. The only evidence of completed payment is an undated memorandum. Longwood Manuscripts, Group 3, #2368.

27 See appendices A and C.

28 Accession 500, Ledger, #934, folios 16, 17.

29 The hyphenated name distinguished this elder brother from his more famous sibling, the French Finance Minister, Jacques Necker. He was also referred to as Mr. Germany at times. Necker-Germany's signature appeared on the Deed of Association but, as mentioned previously, it was obviously inserted almost as an afterthought. This accounts for the nineteen share allocation. See appendix A.

30 "Mr. Necker-Germany . . . seemed willing to take two shares in the Powder and therefore four in the DuPont Company. Mr. Bidermann offered to attend to this subscription." "Notes concerning the Powder Manufacture," undated. DuPont, *Life*, V:245.

31 In a letter from Victor dated August 8, 1801, fully three months after the date of the original Deed of Association, he indicated that "Mr. Necker-Germany will take two shares altogether, not more, and he will take one of them in the powder, if we choose." Victor DuPont to DuPont de Nemours Father, Sons & Company, August 8, 1801. DuPont, *Life*, V:260. He still hadn't made a decision as of August 13, 1801 and in another letter from Victor, he stated that "Mr. Germany is willing to take one but wants only that one . . . and I think I will give it to him." Victor DuPont to E. I. DuPont, August 13, 1801. DuPont, *Life*, V:266.

32 According to the first record book, Necker-Germany paid DuPont de Nemours Father, Sons & Company "in full" for his one share investment in the gunpowder manufactory. Accession 500, Journal, #877, first page, unnumbered. Accession 500, Ledger, #934, folios 20, 21. He may also have agreed to finance one share on behalf of Pierre's firm. The timing and completion of the promised payments was not clear since such payments did not show up in the record books attributed to Necker-Germany and Pierre's records were woefully inadequate. See, for example, Longwood Manuscripts, Group 1, #470 and #478. The only

evidence of completed payment is an undated memorandum. Longwood Manuscripts, Group 3, #2368. In any event, Necker-Germany's refusal to adhere to the two-for-one share purchase policy meant that the parent firm could not even claim to fund as many shares as had been planned.

33 The date of his death is unknown. His survivors were "Mr. Necker of Saussure and . . . Madame Rillet-Necker, his son and daughter." Renewal of Gunpowder Manufactory, May 25, 1808. DuPont, *Life*, VIII:73.

34 Accession 500, Ledger #934, folios 20, 21.

35 Victor DuPont to DuPont de Nemours Father, Sons & Company, August 8, 1801. DuPont, *Life*, V:257-8.

36 The only direct evidence of this is an undated memorandum. Longwood Manuscripts, Group 3, #2368.

37 On at least two occasions, Pierre asked E. I. to mortgage the gunpowder manufactory to repay the loan. On April 24, 1802 he asked E. I. to mortgage the gunpowder manufactory to Necker for "9000 dollars." P. S. DuPont to E. I. DuPont, April 24, 1802. DuPont, *Life*, VI:25. Again on April 27, he asked his son to secure a "mortgage on the manufacture in Mr. Necker's favor" for "8889 dollars 65 cents." P. S. DuPont to E. I. DuPont, April 27, 1802. DuPont, *Life*, VI:31. There are no indications that he ever did so, however, and no Bills Payable for either of these amounts appeared in the Ledger. Accession 500, Ledger, #934, folio 4.

38 Accession 360. Auguste de Staël to E. I. DuPont, March 1, 1813. DuPont, *Life*, IX:90. Longwood Manuscripts, Group 1, #526.

39 Necker's loan "formed the principal part of the capital furnished by the Company DuPont Pere for the company," although Necker never owned any shares directly in the gunpowder manufactory. "Notes written for the information of Du Pont de Nemours," undated. DuPont, *Life*, X:39.

40 Accession 500, Journal, #877, first page, unnumbered. The entry was undated. The Ledger entry was also undated although the entry recognizing a receivable by debiting McCall's Account Proper for the amount of his investment was dated June 1, 1802. Accession 500,

Journal, #877, folio 2. Accession 500, Ledger, #934, folio 23.

41   See Appendix A. His name appeared on this document even though his investment occurred some time after the document was drafted.

42   Victor DuPont to E. I. DuPont, August 17, 1802. DuPont, _Life_, VI:102. McCall also offered the struggling gunpowder manufactory a much needed source of funds. As E. I. noted "Mr. McCall, whose credit in Philadelphia may be useful to us in purchasing materials, has offered me from 4000 to 5000 Ds." E. I. DuPont to P. S. DuPont, August 15, 1802. DuPont, _Life_, VI:96.

43   Victor DuPont to E. I. DuPont, September 28, 1802. DuPont, _Life_, VI:119. In October of the same year, Victor indicated "my advice about McCall is that it is best to give him the three shares, but on the express condition that he pays immediately." Victor DuPont to E. I. DuPont, October 26, 1802. DuPont, _Life_, VI:129.

44   Accession 500, Journal, #877, folio 2. Accession 500, Ledger, #934, folio 23.

45   E. I. DuPont to Archibald McCall, September 5, 1803. DuPont, _Life_, VI:274-5. Accession 500, Journal, #877, second page, unnumbered and folio 2. Accession 500, Ledger, #934, folios 23, 24.

46   Accession 500, Ledger, #934, folios 23, 33.

47   E. I. DuPont to Peter Bauduy, November 30, 1801. DuPont, _Life_, V:313.

48   Holland, _Families_, p. 27. E. I. DuPont to William Hamon, April 26, 1802. DuPont, _Life_, VI:29-30.

49   Accession 500, Journal, #877, second page, unnumbered. Accession 500, Ledger, #934, folios 25, 26.

50   "The certain loss of all my property in San Domingo changes all my plans in this country; and I can no longer consider having an interest in your manufacture on the Brandywine." William Hamon to E. I. DuPont, September 8, 1803. DuPont, _Life_, VI:275.

51   Accession 500, Journal, #877, second page, unnumbered. Accession 500, Ledger, #934, folio 25.

52 "Take your time, my friend, about returning the
    $1500." William Hamon to E. I. DuPont, September 8,
    1803. DuPont, Life, VI:276. His original investment
    was repaid in March 1804. Accession 500, Journal,
    #877, folio 6.

53 See appendix A.

54 Peter Bauduy to E. I. DuPont, October 5, [1801].
    DuPont, Life, V:289.

55 E. I. DuPont to Peter Bauduy, June 25, 1802. DuPont,
    Life, VI:73.

56 "Propositions made by Peter Bauduy to Messrs. du
    Pont," undated. DuPont, Life, VI:74-5. "E. I. du
    Pont's answer to the foregoing," undated. DuPont,
    Life, VI:76-80.

57 E. I. DuPont to P. S. DuPont, August 15, 1802.
    DuPont, Life, VI:96-7.

58 Victor DuPont to E. I. DuPont, August 12, 1802.
    DuPont, Life, VI:93-5.

59 Articles of Agreement with Peter Bauduy, August 25,
    1802. DuPont, Life, VI:108-9.

60 See Appendix A. Bauduy may have signed the original
    Deed of Association at this time, well after the date
    of the document.

61 Accession 500, Journal, #877, second page, unnumbered.

62 Early in the association Bauduy apparently fulfilled
    the requirements of his investment by paying some of
    the debts of the firm as they were incurred. Selected
    line items beginning in August 1802, when Bauduy
    officially affiliated with the gunpowder manufactory,
    were annotated with the letters "PB" in the Factory
    Building Book. These amounts were included as
    expenses of the manufactory when the information in
    the Factory Building Book was transferred to the
    Journal in February 1806. Accession 500, Blotter,
    #848, folios 6-46. Accession 500, Journal, #877,
    folio 71.

63 See appendix A, Article 4 of the Deed of Association.

64 Charles E. Freedeman, Joint-Stock Enterprise in France
    (Chapel Hill, North Carolina: The University of North
    Carolina Press, 1979), pp. 3-18.

65 See Longwood Manuscripts, Group 1, #477, for a copy of the prospectus with the phrase "societé en commandité" in the description of the firm.

66 The DuPonts discovered that unlimited liability did not exist at this time in America only after their arrival. "Statement made by DuPont (de Nemours) Father to the Shareholders in his Company," April 18, 1808. DuPont, Life, VIII:49.

67 McCall had placed the advertisement apparently with the knowledge and approval of Bauduy, but without the prior authorization by E. I. E. I. DuPont to Victor DuPont, [December 13, 1804]. DuPont, VII:50-3. Misunderstandings on both sides fueled the disagreement. A vitriolic exchange erupted between the two men with charges and countercharges on both sides. E. I. suspected that Bauduy was trying to usurp his position as founder and Director of the firm, and relegate him to the position of head powderman. Bauduy, on the other hand, believed that E. I. was trying to present him as only a company clerk and not as a significant investor in the gunpowder manufactory. The conflict became so bitter that at one point Bauduy threatened to withdraw from the firm. Peter Bauduy to Victor DuPont, December 13, 1804. DuPont, VII:68-77. If he had taken such drastic action, the financial base of the enterprise would have been removed as well. He by this time owned four shares in the gunpowder manufactory, and had arranged and funded a number of loans for the firm. The family could ill afford such a financial crisis, and was in no position to redeem Bauduy's various investments. Finally, Victor stepped in as intermediary to quell the exchange. Victor DuPont to E. I. DuPont, December 17, 1804. DuPont, Life, VII:89-97. He was moderately successful, and eventually the controversy calmed.

68 "Copy of secondary agreement with Mr. Bauduy." DuPont, Life, VII:146-8. This move, although apparently satisfactory to Pierre at least in the short run, did not serve to allay E. I.'s suspicions or concerns with respect to his partner. He continued to profess uneasiness and even fear concerning Bauduy's possible actions. At best, E. I. felt that Bauduy would not do the firm any harm, although he still expressed concern over the continued partnership. E. I. DuPont to P. S. DuPont, April 12, 1808. DuPont, Life, VIII:38.

69 See appendix A, Articles 13, 14 & 15. The parent
    company's sizable ownership interest as well as the
    number of other shareholders friendly to the family
    virtually ensured continuation of the partnership.

70 Accession 500, Journal, #878, folio 51 & 55.

71 Accession 500, Journal, #878, folio 54.

72 Accession 500, Ledger, #934, folio 17 & 19. The
    entries were recorded in the appropriate Ledger
    accounts although no corresponding Journal entry was
    made at this time.

73 E. I. DuPont to P. S. DuPont, May 26, 1811. DuPont,
    Life, VIII:302.

74 Accession 500, Waste Book, #862, folio 4. The
    allocation of shares on January 1, 1810 did not occur
    without some ill feeling on the part of Pierre,
    however. Apparently, E. I. made a separate agreement
    with Bauduy on this date, and in some way excluded his
    father's firm. A copy of this agreement has not been
    located. On May 25, 1808, Pierre, as representative
    of DuPont de Nemours Father, Sons & Company, initiated
    a document authorizing renewal of the gunpowder
    manufactory. Untitled, May 25, 1808. DuPont, Life,
    VIII:72-3. This authorization must have reached E.
    I., although his recognition of this fact apparently
    did not, in turn, reach his father. The mail was very
    bad, however, and letters often did not get through.
    Consequently, although Pierre apparently
    misunderstood, the company had indeed been renewed
    with all shares allocated as required. E. I. DuPont
    to P. S. DuPont, May 26, 1811. DuPont, Life,
    VIII:302. P. S. DuPont to E. I. and Victor DuPont,
    October 14, 1813. DuPont, Life, IX:119.

75 "Statement made by DuPont de Nemours Father to the
    Shareholders in his Company," April 18, 1808. DuPont,
    Life, VIII:63. In this document the date of
    dissolution is set at July 1812. In another letter,
    however, the date is given as July 1811, perhaps an
    error made when the letter was copied. "Copy of a
    letter from P. S. DuPont to E. I. DuPont," undated.
    DuPont, Life, VIII:119.

76 Accession 360. Auguste de Staël to E. I. DuPont,
    March 1, 1813. DuPont, Life, IX:90.

77 Longwood Manuscripts, Group 1, #526. "Statement made
    by DuPont (de Nemours) Father to the Shareholders in

his Company," April 18, 1808. DuPont, Life, VIII:40-66. "Sequel and Summary of the statement given April 18, 1808, to the Shareholders of the Company," undated. DuPont, Life, VIII:304-5. Auguste de Staël to E. I. DuPont, March 1, 1813. DuPont, Life, IX:90.

78 E. I. DuPont to P. S. DuPont, November 1, 1812. DuPont, Life, IX:56-61. E. I. was very concerned by the distribution his father had made. He felt that some of the shares might "fall in the hands of strangers, perhaps enemies," specifically Bauduy. E. I. DuPont to P. S. DuPont, December 10, 1811. DuPont, Life, IX:18-19. In addition, Pierre had also secured a loan from Talleyrand through his agent, Menestrier, for 100,000 francs in 1807. This obligation which was to be repaid in 1813 also became E. I.'s responsibility since the proceeds had been used to some extent by Pierre to subsidize his firm's investment in the gunpowder manufactory. On December 31, 1815, the record keepers recorded "received of him [Menestrier] on the 6th Novr 1807 by P. S. DuPont for our account $18500. Accession 500, Journal, #879, folio 337. The account was carried in Menestrier's name throughout the period under consideration. See Accession 500, Ledger, #935, folio 257 and #936, folio 120.

79 E. I., apparently acting on his negative feelings regarding Bauduy, continued to plot a way to remove him from the partnership. His father, Pierre, warned against such a move, however. "Copy of a letter from P. S. DuPont to E. I. DuPont," undated. DuPont, Life, VIII:117. Bauduy, although a trial, was better as a lukewarm partner, than as an enemy, he warned. Peter Bauduy, for his part, apparently decided to continue his affiliation with the firm. Apparently his alienation did not extend to giving up his interest in what to him seemed a potentially profitable venture.

80 Peter Bauduy to E. I. DuPont, postmarked January 23, 1814. DuPont, Life, IX:168. Guide, p. 280.

81 "Copy of a letter written by Mr. Bauduy to Mr. Beauchet on June 21, 1814, and sent to Mr. Bidermann," undated. DuPont, Life, IX:196-200.

82 Bauduy sold his interest in the firm for $60,000. Accession 500, Waste Book, #862, folio 254. Accession 500, Journal, #879, folio 276.

83 Accession 500, Journal, #879, folio 277.

84   Accession 146, File 51.

85   Many of the account books from this period were
     annotated as part of the evidence in the "7th
     Interrogatory." Examples -- Accession 500, Journal,
     #877, inside front cover.   Accession 500, Ledger,
     #934, inside front cover.

Chapter Three

ACCOUNTING FOR OPERATIONAL CASH FLOWS,
CASH FLOW DEFICIENCIES, AND CREDIT RECEIVABLES AND PAYABLES

As is evident, the capitalization of the DuPont Company resulted in little direct cash investment. Further, the firm's decision makers participated in very few direct cash transactions. This lack of cash was not an unusual circumstance during this period, however, and DuPont record keepers adapted the company bookkeeping to document the methods devised by market participants to avoid the transfer of cash. The bookkeeping used to accomplish this documentation was straightforward and direct and record keepers familiar with bookkeeping techniques easily accommodated the remarkably pervasive and efficient credit-based system which existed. The success of this system allowed firms to easily devise alternative means of fulfilling promises to pay acceptable to all parties concerned. DuPont Company record keepers therefore documented both formal and routine credit transactions and occasional bartered settlements of accounts in addition to the very few direct cash transactions which occurred.

Record keepers debited the "Cash" account to indicate cash receipts, and credited the same account for cash disbursements. Most of these cash transactions related to existing credit arrangements, however. Company bookkeepers designated these notes or drafts "Bills Receivable" or "Bills Payable" depending on the originator of the obligation, and each note or draft specified a duration, amount and required interest charge.[1] In addition, formal debt instruments could be substituted for routine obligations.[2]

60

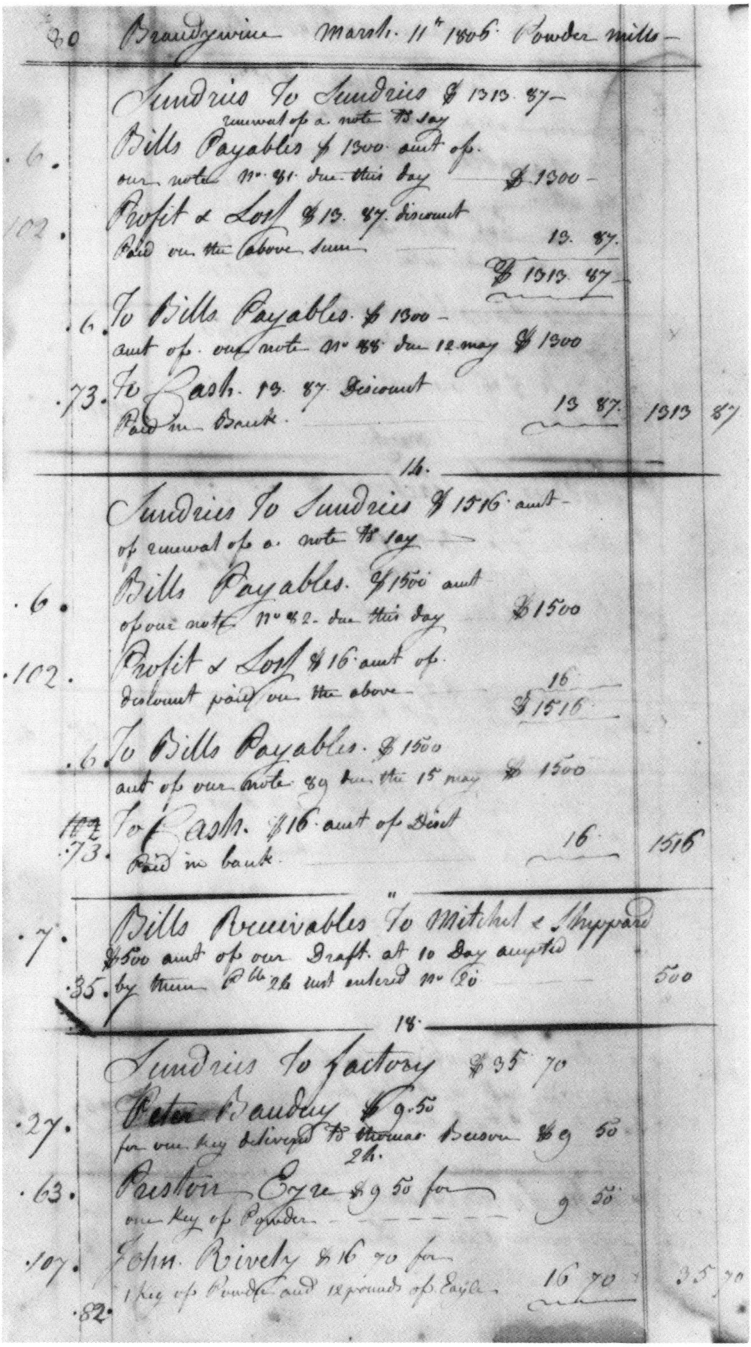
80 Brandywine March 11th 1806 Powder mills

Sundries To Sundries $ 1313. 87
renewal of a note $ say
.6. Bills Payables $ 1300 amt of our note No. 81 due this day $ 1300
.102. Profit & Loss $ 13. 87 discount paid on the above sum 13. 87
$ 1313. 87
.6. To Bills Payables $ 1300 amt of our note No. 88 due 12 may $ 1300
.73. To Cash $ 13. 87 Discount paid in Bank 13 87 1313 87

14

Sundries To Sundries $ 1516 amt of renewal of a note $ say
.6. Bills Payables $ 1500 amt of our note No. 82 due this day $ 1500
.102. Profit & Loss $ 16 amt of discount paid on the above 16
$ 1516
.6. To Bills Payables $ 1500 amt of our note 89 due the 15 may $ 1500
.73. To Cash $ 16 amt of Disct paid in bank 16 1516

"

.7. Bills Receivables To Mitchel & Sheppard $500 amt of our Draft at 10 Day accepted
.35. by them Bl 26 not entered No 20 500

18

Sundries To Factory $ 35 70
.27. Peter Baudey $ 9.50 for one key delivered to Thomas Benson 2h $ 9 50
.63. Preston Eyre $ 9 50 for one key of Powder 9 50
.107. John Rively $ 16 70 for 1 key of Powder and 12 pounds of Eagle 16 70 35 70
.82.

ILLUSTRATION 6. Examples of Bills Payable, Bills Receivable and other routine credit transactions. Accession 500, Journal, #877, folio 80. Photograph courtesy of Hagley Museum and Library.

The record keepers debited the "Bills Receivable" account to document the existence of a credit instrument owed to the firm. The following entry, dated March 14, 1806, depicted a situation where a short-term draft was sent to a firm for payment on March 24, 1806.

```
.7.      Bills Receivable to Mitchell & Sheppard
         $500 amount of our Draft at 10 days accepted
   .35. by them payable 24 instant [March 24] entered
         No. 20                                    500    3
```

Cash payments received for both formal and routine receivables were debited to the "Cash" account and the receivables accounts were credited. Both of these situations were presented in the following entry dated March 24, 1806.

```
.73.     Cash to Sundries              $548
    .7.  To Bills Receivables $500 amount
         of received in bank and put to the
         credit of E. I. D. P./bill No. 20   $500
   .63. To Preston Eyre $48 amount
         of cash paid to E. I. D. P.          48   548   4
```

The bookkeepers documented the liabilities of the Company by crediting the "Bills Payable" account for the amount of any formal obligation owed by the firm. The offsetting debit(s) recorded the purchase of raw materials, the acquisition and construction of operating facilities, cash inflows from short-term notes, and other miscellaneous events. Thus, formal credit instruments constituted an important source of funding for the early DuPont Company. The first "Bill Payable" was issued on October 25, 1803, and cash was received. The following entry documents this event.

            Sundries to Bills Payables $1500 amount of
            our note No. 1 due the 27 December to say
    2.   Cash $1484 net proceeds $1484
    12.   Profit & Loss $16 discount on $1500   16   1500
        4.   5

The following year, the company issued over thirty-five notes.[6] More were issued in subsequent years. The notes did not ensure the inflow of a large amount of cash, however, primarily because most of the transactions subsequent to those occurring in the very beginning involved the renewal of an existing obligation under the guise of a new note with a new number. Thus, when "Bills Payable" No. 1 was due for collection, No. 2 was issued, and the following entry was made on December 27, 1803.

            Sundries to Sundries $3000 as follows

    2.   Cash $1484 net proceeds of $1500 $1484
    12.   Profit & Loss $16 discount of ditto   16
    4.   Bills Payables $1500 Note No. 1
                    to take up   1500
                                                   3000
            4. To Bills Payable $1500 amount
            of our Note No. 2 due the 28 february 1500
            2. To Cash $1500 put in Bank   1500 3000   7

In this way a fairly moderate balance appeared in the "Bills Payable" account throughout most of the period under consideration.[8]

The obligations of the DuPont Company could also be affected by the "acceptance" of the credit obligations of another party for payment. As an example of an acceptance, on February 24, 1804, the DuPont Company assumed responsibility for specific obligations of Victor DuPont de Nemours & Company which became a "Bills Payable" of the gunpowder manufactory. The company record keepers entered the transaction as follows.

```
        V. DuPont de Nemours & Co. to Bills Payables
22.            $8640.40 amount of their drafts
  4.               for 4300 in favor Governor & Kemble destroyed
                4340.40 and replaced by other drafts

No. I      $2019.56     II      $2340.40
   I bis    2280.44     II bis   2000
                    due 25 May 4300 accepted 8640.40      9
```

In addition to formal credit arrangements company customers, agents and even principals participated in many routine credit transactions. It was possible that a merchant who sold supplies or raw materials to the gunpowder manufactory and also interacted with the firm in another capacity as customer, agent or principal was assigned several different accounts including amounts owed by the gunpowder manufactory to the merchant and by the merchant to the DuPont Company.[10] No matter which category an individual or firm belonged to, however, company record keepers debited the associated accounts to indicate receivables and credited them to indicate payables.

Following the innovation in the company record keeping of February 11, 1806, the accounts created in the ledgers relating to company customers, agents and principals included information on receivables, payables, and payments on account. The only way to determine whether an account constituted a receivable or payable was to identify the balance in that account. If the debit total was greater than the credit total, the account functioned as a receivable. If the reverse was true, however, the DuPont Company had incurred a payable equal to the credit balance.

Most of the transactions documented by the record keepers concerned the disposition of the company's product. A credit sale to a routine customer was debited to an account under his name. Sales to company principals were debited to their "Account Proper." The following entry, dated March 18 & 24, 1806, depicted both of these situations.

```
                        18
              Sundries to Factory   $35.70
  .27.   Peter Bauduy $9.50
         for one keg delivered to Thomas
         Beeson                                  $9.50
                        24
  .63.   Preston Eyre $9.50 for
         one keg of powder                       9.50
  .107.  John Rively $16.70 for
         1 keg of powder and
         12 pounds of Eagle                     16.70    35.70
     .82.                                                        11
```

The account of an agent was annotated with the word "Factorage" in the early account books. An entry dated March 28, 1806, indicating the delivery of powder to an agent, Anthony Girard, and the recognition of a receivable was therefore recorded as follows.

```
         Anthony Girard factorage to factory $2570
  119.   Amount of 48 kegs invoiced to him,
         40 kegs at 9.50. 8 kegs at 12.50         $480
    82.  Amount of 220 kegs invoiced to him at
               9.50                              2090 2570   12
```

Credits to the accounts of customers, agents or company principals indicated amounts owed to these individuals or firms. Payables to customers or agents resulted from transactions involving the operational expenses of the manufactory and the purchase of raw materials.[13] Payables to company principals resulted from similar transactions if they applied, or at least interest on their investment, as well as the potential return of the investment itself.[14]

The workers occupied a separate category with respect to the cash flows and routine credit transactions of the DuPont Company. Information concerning specific workers was maintained in the company "Petit Ledgers." The record keepers recorded both cash payments and barter arrangements whereby the firm provided or paid for goods, services or other eventualities. The payroll constituted a drain on cash resources which could be deferred when barter arrangements were accepted by the workers in payment.[15]

The fact that such organized credit opportunities existed during the early nineteenth century implied that cash flow deficiencies were not a unique occurrence during this period. The procedures used by the company record keepers to document this largely credit-based society were straightforward and not particularly innovative. In this way, little cash changed hands and rarely can a relationship be established between a credit transaction and an associated collection. The recognition of the more encompassing concepts of revenues, expenses and profits or losses resulted in some rather unexpected developments, however, which will be presented in the following chapter.

Notes--Chapter Three

1   The standard credit entry concerned a sixty-four day
     note with a six-percent interest charge.

2   Record keepers took information that had been debited
     to a personal account to indicate a receivable or
     credited to the account to indicate a payable, and
     placed it in the "Bills Receivable" or "Bills Payable"
     account.   See, for example, the entry for Joseph
     Sumert dated June 29, 1804.   Accession 500, Journal,
     #877, folio 10.

3   Accession 500, Journal, #877, folio 80.

4   Accession 500, Journal, #877, folio 81.

5   Accession 500, Journal, #877, folio 3.

6   Accession 500, Ledger, #934, folios 4 & 5.

7   Accession 500, Journal, #877, folio 4.

8   The balance ranged between $3000-$10000 for most of
     this period.   Large amounts recorded in the last two
     years were due to significant purchases of saltpetre
     on credit.   Accession 500, Ledger, #936, folio 42.

9   Accession 500, Journal, #877, folio 5.

10   See, for example, the accounts for Archibald McCall.
     Accession 500, Ledger, #934, folio 23--Account Proper;
     folio 24--Account in Company; folio 33--Factorage
     Account.

11   Accession 500, Journal, #877, folio 80.   Folio 27 was
     Bauduy's Account Proper.   Eagle was a special kind of
     powder.

12   Accession 500, Journal, #877, folio 81.

13   See, for example, the entry for Archibald McCall dated
     February 15, 1810.   Accession 500, Journal, #878,
     folio 65.

14   See, for example, the entries for company principals
     DuPont de Nemours Father, Sons & Company, Duquesnoy,

Necker-Germany, E. I., and Bauduy.   Accession 500, Journal, #878, folio 51 & 52.

15   Accession 500, Blotter, #849.   Examples of the kinds of goods or services provided by the company in lieu of cash are given in the "Annotated Bibliography."

Chapter Four

REVENUES, EXPENSES, PROFITS AND LOSSES

The credit-based economy of the early nineteenth century essentially forced record keepers of the period to maintain the record books using a modified cash basis. The bookkeepers could not rely on cash flow information alone to identify the firm's economic position. Thus, as a result of cash and credit transactions involving agents, customers and company principals the DuPont Company record keepers essentially recognized the expanded concept of revenues earned and expenses incurred in the operations of the gunpowder manufactory, regardless of when the associated cash was paid or received. The bookkeepers were then able to identify the financial position of the firm by isolating the profit or loss for a particular span of time.

The bookkeeping treatment for revenues earned as a result of gunpowder sales depended on the nature of the sale. Factory-direct cash or credit sales to customers or company principals were rare. When they occurred, such direct purchasers agreed to pay the price charged at the manufactory for the powder.[1] E. I. effected most sales through company agents, however, so that DuPont gunpowder would be widely distributed geographically. He offered company agents two alternatives for procuring the powder.[2] In one case the agent could "buy" the powder outright and then remit the amount agreed to by the two parties to the firm after he sold the powder. This simply meant that the agent was charged a particular invoice price and he in turn charged the customers for that amount plus any profit he was able to make and whatever other expenses he incurred. As an alternative the powder was not sold to the agent but rather was consigned to him for sale on a commission basis. Generally, under this method, the agent sold the powder for a price and owed the firm that amount less a commission and other expenses as agreed to by the two parties.[3] In both cases, the treatment of warehousing and transportation costs varied depending on the amount of the sale, the location of the purchaser, and other circumstances as they arose. These additional expenses relating to gunpowder sales were not detailed separately, but were either included in the invoice or sales price of the goods, or charged to the recipient of the property.[4] Particularly in the earliest days, E. I. and other decision makers maintained control, as much as possible, of the selling price of the powder.[5] Eventually, however, they recognized that some measure of discretion had to be left to the agents.[6]

Initially, the record keepers credited all powder distributions directly to the "Factory" account.[7] For direct sales to customers or company principals, this constituted an accurate recognition of revenue from a completed earnings process regardless of when the associated cash collection or fulfilled receivable occurred. Transfers of powder to agents for disposal on behalf of the company did not constitute a completed earnings process, however, and therefore should not have been credited to the "Factory" account until the agents sold the powder to a third party.[8]

The "Factory" account included more than just this revenue information, however. The record keepers also debited the "Factory" account for general operating expenses incurred in the operations of the business, specifically for the cost of using raw materials to manufacture the powder. The nineteenth-century record keepers did not call this "Cost of Goods Manufactured," as we would today. Nonetheless, when they debited the

70

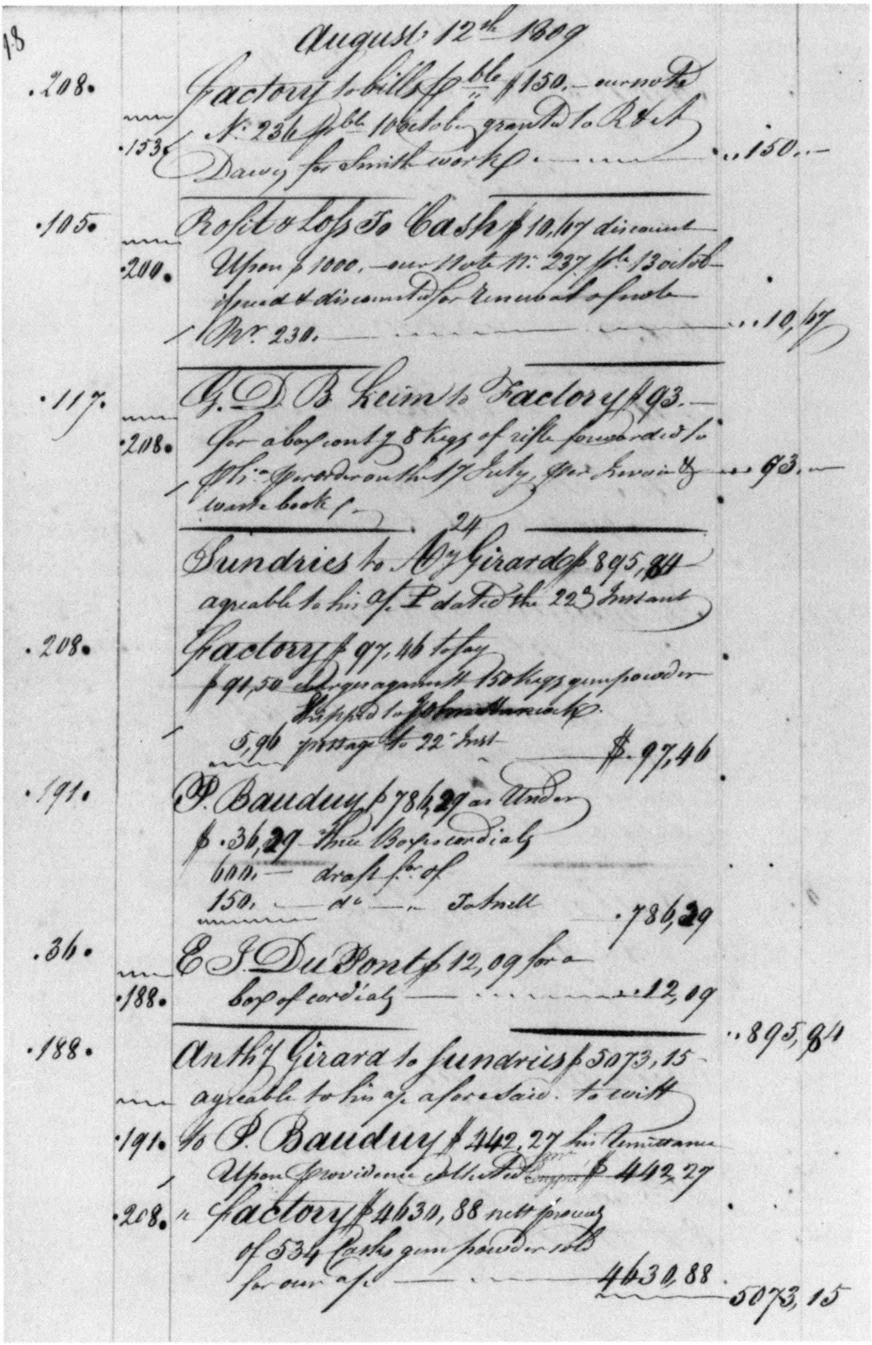

18

August 12th 1809

.208. Factory to bills payable $150.– our note
.153. No. 236 payable 10 October granted to Robt
Davey for Smith work ... 150.–

.105. Profit & Loss To Cash $10,67 discount
.200. Upon $1000. our Note No. 237 payable 13 octob
issued & discounted for renewal of note
No. 230. ... 10,67

.117. G. D. B. Keim to Factory $93.–
.208. For a box cont'g 8 kegs of rifle forwarded to
Wm. ... on the 17 July, per invoice & ... 93.–
ware book

24
Sundries to Anty Girard $895,84
agreeable to his a/c dated the 22d instant
.208. Factory $97,46 to pay
$91,50 charges against 150 kegs gunpowder
shipped to ...
5,96 postage to 22 inst ... $97,46
.191. P. Bauduy $784,29 as Under
$36,29 three Boxes cordials
600. draft for of
150. do Tatnell ... 786,29
.36. E. I. Du Pont $12,09 for a
.188. box of cordials ... 12,09 ... 895,84

.188. Anthy Girard to Sundries $5073,15
agreeable to his a/c aforesaid. to wit
.191. To P. Bauduy $442,27 his Remittance
Upon Providence collected ... $442,27
.208. " Factory $4630,88 nett proceeds
of 534 Casks gunpowder sold
for our a/c ... 4630,88 ... 5073,15

ILLUSTRATION 7. Examples of entries involving the "Factory" account. Accession 500, Journal, #878, folio 18. Photograph courtesy of Hagley Museum and Library.

"Factory" account and credited the raw materials accounts to indicate the amounts used in the process of manufacturing the gunpowder, they accumulated this information.[9]

Company record keepers apparently recognized that the premature crediting of the "Factory" account had to be adjusted to reflect the ending inventory of gunpowder on hand, or in the hands of agents, at the end of the first partnership. In December 1809, therefore, Duplanty made an entry debiting a new account labelled "Gunpowder" and crediting the "Stock" account.[10] Eventually, the record keepers credited a new account labelled "Gunpowder" for all powder distributions and debited the same account to recognize the actual sales of powder. The offsetting credit to the "Factory" account occurred at this point. Ending inventory was debited to a "Gunpowder Outstanding" account, and included powder "in the hands of several agents."[11]

In addition to debiting the "Factory" account for operating expenses, the record keepers also debited the "Profit and Loss" account. The "Profit and Loss" account included special expense information such as interest charges on renewed or retired notes and interest on invested capital.[12] The record keepers also debited the "Profit and Loss" account on December 31, 1809, for "presumed depreciation in the value of our buildings and machinery in virtue of inventory made this day." They credited the "Factory Buildings" account for the same amount, presumably to recognize the results of the explosion which occurred in 1808.[13] The depreciation entry by name did not recur, although an additional entry to adjust the value of the account occurred at the end of the period under study.

Articles Six through Nine of the 1801 Act of Association stipulated that the profit or loss earned by the firm be identified on an annual basis.[14] Despite this stipulation, the closing process necessary to the identification of the firm's financial position and therefore earned profit or loss occurred only seven times during the period under consideration.[15]

The closing process meant that all debit balance accounts were credited, and the credit balance accounts were debited in the "Trial Balance." The balance account could either have been a separate location in the Ledger or the "Stock" account itself. The "Stock" account, initially debited and credited for the $36,000 of capital investment originally anticipated by E. I., constituted essentially a summation or "Trial Balance" account itself at times. A debit to that account meant a loss, and a credit meant income.

The record keepers of the DuPont Company during the period under study did not need to incorporate extremely sophisticated techniques into the documentation process. They did, however, need to adapt existing procedures to meet the needs of a developing manufacturing firm.

Notes--Chapter Four

1   These straightforward entries usually involved a small
    amount of powder, and were very rare. E. I. generally
    preferred that all customers go through one of his
    agents.

2   In many cases E. I. suggested one of these
    alternatives if specific circumstances
    warranted. See, for example, a letter to Mr. John
    Gundacker. Accession 500, Letter Book, #773, folio
    50, August 27, 1806.

3   A commission expense generally not identified
    separately was recognized for agents at
    two-and-one-half percent of sales and was excluded
    from the price to agents to be remitted to the firm.

4   Depending on the location of the agent, and the
    agreement he had with the DuPont Company,
    transportation, warehousing and insurance costs might
    have been his responsibility.

5   See, for example, a letter to J. D. P. Dawes,
    merchants in Albany. Accession 500, Letter Book,
    #773, folio 40, June 24, 1806.

6   See, for example, a letter to William Cornell of New
    York. Accession 500, Letter Book, #776, folio 114,
    June 22, 1814.

7   See, for example, the entries for McCall and Mitchell
    & Shephard. Accession 500, Journal, #877, folio 16.

8   Further complicating this issue, if the powder was not
    sold, or was sold for a different value than the
    Invoice price because of a change in the market price
    of the powder, the earlier recognition was
    invalidated. Company record keepers never recognized
    this situation, however.

9   The raw materials accounts included saltpetre,
    brimstone, charcoal wood, and the cooperage account
    for barrels and kegs. See, for example, Accession
    500, Journal, #879, folio 25.

10   This entry therefore recorded ending inventory and effectively reduced the value of the costs attributable to the use of raw materials which left what constituted cost of goods sold in the accounts. Accession 500, Journal, #878, folio 60. This information was transferred to the "Gunpowder Outstanding" account in the new Ledger. Accession 500, Ledger, #935, folio 12.

11   Accession 500, Journal, #879, folio 185. The "Gunpowder Outstanding" account had to be adjusted at a later date to reflect the powder which had been sold, and credited to the "Gunpowder" account. Accession 500, Journal, #879, folio 259.

12   The interest allocated to the investors at six-percent was adjusted on December 31, 1810, "because the stockholders wish to be credited with compound interest." Accession 500, Journal, #879, folio 26.

13   Accession 500, Journal, #878, folio 60.

14   See appendix A.

15   February 11, 1806. Accession 500, Journal, #877, folios 74-77. Not a complete Trial Balance.
December 31, 1809. Accession 500, Ledger, #934, folios 224-225. Accession 500, Waste Book, #862, folios 1-4.
June 30, 1814. Accession 500, Ledger, #935, folios 211-212. Accession 500, Journal, #879, folios 208-211.
December 31, 1814. Accession 500, Journal, #879, folios 264-267.
December 31, 1815. Accession 500, Ledger, #935, folios 213-214.
December 31, 1817. Accession 500, Journal, #880, folios 134-137.
August 31, 1818. Accession 500, Journal, #880, folios 236-241.

Chapter Five

CONCLUSION

Accounting as a research discipline tends to focus on current research problems and disregard the rich history which predates the requirements developed from the Securities and Exchange Acts of 1933 and 1934. Such a delimited discipline demands analysis of historical data only as it pertains to contemporary issues. Thus, accountants tend to ignore the fact that the great advances evidenced today are based on historical developments to meet specific needs. The usefulness criterion resulted in adaptations in the practice of accounting to meet changing business environments, constructs, and forms. Basically, however, the fundamental practice of bookkeeping evidenced in the nineteenth century still applies.

In the nineteenth century, firms needed to adapt record keeping procedures to meet existing needs. Usefulness was the primary criterion which had to be accommodated, and the DuPont Company bookkeepers evidenced the realization of this criterion to a surprising degree. Even more surprisingly, the differences between the methods developed at the beginning of the last century to accommodate certain needs and those touted as incredible advances today do not seem very significant. Allowing for some of the limitations of the time, the differences are a matter of form instead of substance.

From the beginning, the DuPont Company record keepers had to adapt the bookkeeping practice that had developed to accommodate the needs of merchandising concerns to a manufacturing entity. The basic practices stayed the same, but the record keepers did have to somehow address the

76

issue of raw materials used in the production process. They did this quite effectively. These nineteenth-century record keepers even went so far as to value what constituted "Work-in-Process Inventory," and recognized completion of this in-process powder in ensuing periods.[1]

Further, although there is no clear indication that the company record keepers or decision makers calculated the "cost" of the powder sold, aside from the valuation process recognized in the preceding paragraph, E. I. did recognize that fluctuations in the price of at least one raw material, saltpetre, influenced the minimum amount that could be charged for the final product.[2] They must have had some concept of the underlying costs, therefore, even though there is no clear indication of this fact. This is a fairly sophisticated concept for this period, just before the Industrial Revolution. It is certainly one that still challenges accountants to this day. The cost components have grown that are being considered, but the concept of an underlying "cost" is still at issue.

The valuation of ending inventories of raw materials, finished, and unfinished powder occurred when a "Trial Balance" was generated. The seven times this happened between 1801 and 1818 cannot possibly be considered "periodically" as modern day accountants expect from an organization. The early DuPont Company functioned in a different time, however. The production of these reports on an annual basis would probably be almost impossible. The company letter books are filled with requests sent to agents for information necessary to "close the books."[3] The uncertainty of the mails, the economic health of the agents in the field, and the market for gunpowder all served to make day-to-day operations difficult enough without aggravating the situation by tying up the decision makers in the firm with a closing process that often was

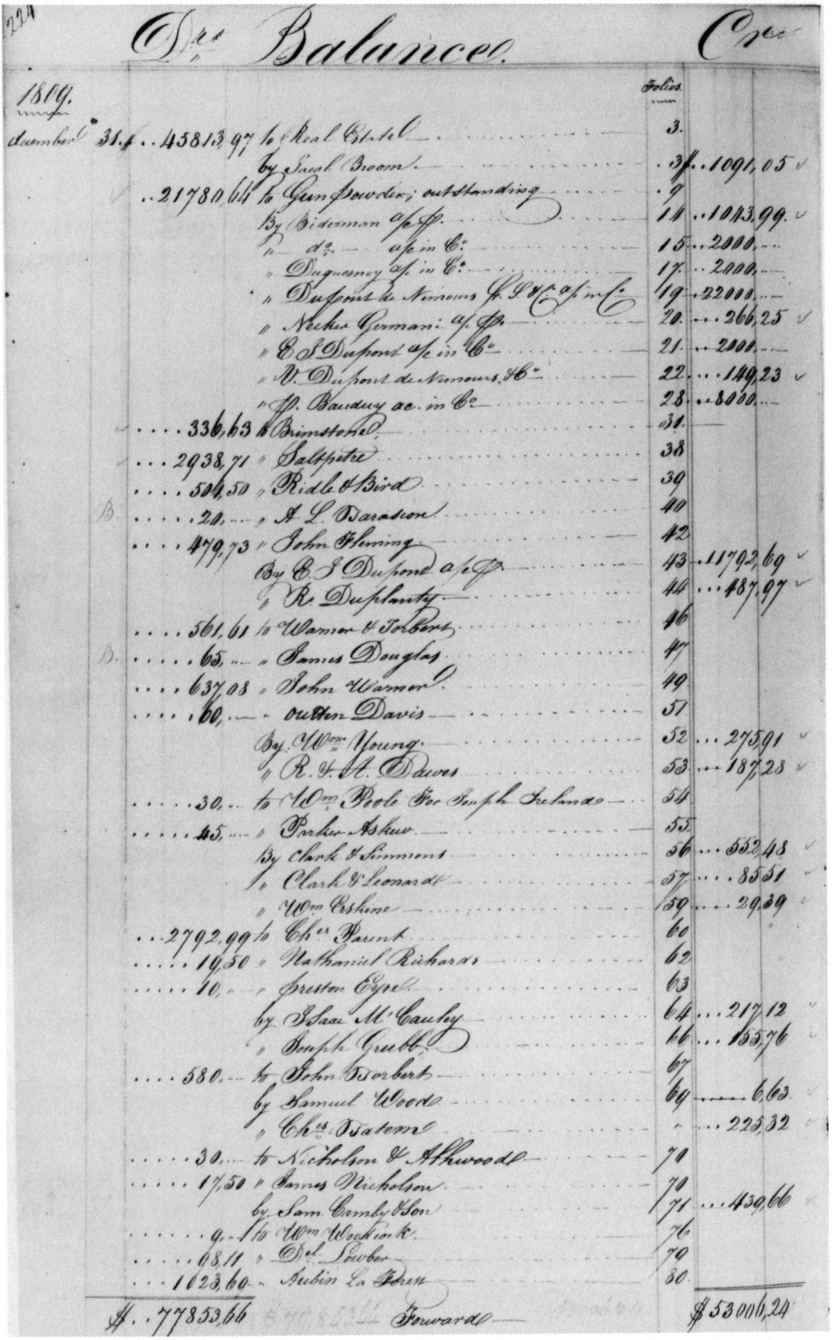

ILLUSTRATION 8. The first page of the "Trial Balance" of December 31, 1809. Accession 500, Ledger, #934, folio 224. Photograph courtesy of Hagley Museum and Library.

not completed for several months. The usefulness criterion would justify the approach taken.

The depreciation entry of December 31, 1809, was used to value the operational assets of the business, in part because of the tendency of this particular enterprise to blow up from time-to-time. The firm never wrote assets up above the costs to build the facilities, but instead recognized that the value of the assets had decreased, because of "ware and tare."[4] In this case, the firm was basically recognizing the true meaning of depreciation, loss of value, and very effectively documenting the fact that the assets involved had less value to the firm at the point that the "Trial Balance" was created. Although nineteenth-century bookkeeping did not take on the sophisticated, periodic allocation of the original value of a particular asset over its estimated useful life, the DuPont Company's approach did make the necessary adjustment to the asset account based on a very real loss of value.

One of the more important results of this research is the recognition that the earliest bookkeeping records of the DuPont Company did not necessarily reflect the underlying reality with respect to dates, dollar values, and events. Thus, when I came across essentially a reference book of basic information or "Statistics" on the company prepared during the mid-1950's as part of an anti-trust suit, I discovered that the DuPont Company Legal Department displayed purported financial statements for these earliest years. The first "Balance Sheet" was dated April 21, 1801, and indicated full capitalization based entirely on the first entry in the Journal.[5] The fact that this was not the case cannot be gleaned from a cursory examination of the extant primary materials.

This research constitutes an historical case study of a particular firm and the accounting procedures adopted to deal with a unique set of circumstances. The techniques used in this case when no mandated or even suggested practices existed for a manufacturing firm cannot be generalized to apply to any other firms of this period. A researcher may seek similarities, but the techniques used in firms of this time must, essentially, be considered firm specific unless some common decision maker influenced more than one entity.

The DuPont Company remained viable during the nineteenth century despite an almost overwhelming set of circumstances. The record keeping practices of the firm, particularly after Duplanty introduced the essentials of double-entry bookkeeping, played an important role in the company's success. The usefulness criterion that caused record keepers to adapt existing procedures to specific needs fostered a valuable set of bookkeeping guidelines in this firm that ensured the success of the company into the following century.

Notes--Chapter Five

1   When the accounts were prepared for the "Trial
     Balance" on June 30, 1814, Duplanty made an entry for
     "G P unfinished in the mills @ 48 Cts" followed
     immediately by an entry "in the magazine ready pick up
     @ 52." Accession 500, Journal, #879, folio 203.

2   Therefore, for example, in a letter from E. I. to
     Dudley & Walsh, company merchants in Albany, he
     indicated that he wished "to keep our prices as low as
     the price of saltpetre can afford us." March 18,
     1807. Accession 500, Letter Book, #773, folio 128.

3   A constant refrain throughout this period is evidenced
     in a letter from the firm to William Cornell in New
     York, where E. I. indicated that he wanted Cornell's
     account of sales, "our accounts being left open for
     want of that document." September 28, 1814.
     Accession 500, Letter Book, #776, folio 140.

4   Accession 500, Journal, #878, folio 60.

5   Accession 1729, DuPont Company Legal Department
     Records, pp. 2-1 to 2-9.

Appendix A

THE DEED OF ASSOCIATION

A Deed of Association relative to the Establishment
of a Manufactory of Gunpowder in the United
States of America

The undersigned [Du Pont de Nemours Father, Sons and
Co. of New York; Bidermann; Catoire, Duquesnoy and Co.; and
Eleuthère Irenée Dupont] wishing to establish a gunpowder
manufactory in the United States of America have entered
into partnership for that purpose, and have agreed as
follows.

ART. 1$\underline{st}$

The Stock of the Company shall amount to thirty-six
thousand dollars, and shall be divided into eighteen shares
of two thousand dollars each.

ART. 2$\underline{d}$

The shares shall be subscribed for & paid by:

[
      Bidermann for one share   one share
      Catoire, Duquesnoy and Compy   one share
        Necker-Germany   one share
         Arch^d McCall   one share
                               one share
      Peter Bauduy   one share
                               one share
DuPont de Nemours Father, and Sons and Co. twelve shares
     of New York   ]

ART. 3$\underline{d}$

The shares will bear Interest at the rate of six per
cent.

ART. 4$\underline{th}$

E. I. du Pont is entrusted with the establishing of
the said factory, and superintendence of the same: his
whole time shall be given to it, and as a compensation for
the same, an allowance of eighteen hundred dollars shall be
made to him every year.

ART. 5$\underline{th}$

The requisite buildings and constructions shall be put
up in the course of the year 1801 & during the beginning of
the next year, so that the factory may be in operation in
the summer of 1802.

ART. 6th

At the close of each year, beginning at the end of December 1803, an Inventory shall be made wherein the real and other property of the concerned shall be valued according to the actual price of the same, at the time of the said valuation. Shall be considered as a profit such part of the said valuation as may exceed the amount of the original stock, after having deducted the interest on the same. The house of Dupont de Nemours Father, Sons and Co. of New York--owning the best part of the shares will find one of its members or appoint some other person provided with a power of attorney to assist at the said Inventory.

ART. 7th

The profits or loss, if any, shall be divided in the following manner: eighteen parts to the owners of shares, nine parts to the chief of the manufactory, as his concern in the establishment created by him; and three parts are reserved for one of those who planned the undertaking.

ART. 8th

If it is not necessary to dispose of the three parts stated in Art. 7, they shall be suppressed.

ART. 9th

The Director of the manufactory and the agent appointed by the House of Dupont de Nemours Father, Sons & Co. will determine each year, after examination of the Inventory what is the portion of profit fit to be divided among the concerned.

ART. 10th

The Interest and dividend due to such of the concerned as live in France, shall be paid by such a house in Paris as will correspond to that effect with that of Dupont Father, Sons & Co. of New York.

ART. 11th

The Director of the manufactory will adopt for his accounts the principles adopted by the administration of powder & saltpetre of France.

ART. 12th

In Case the Director happens to die, the House of Dupont de Nemours Father, Sons & Co is authorized to settle his account with the company, to appoint another person to supersede him, and also to provide for all other requisite arrangements rendered necessary by circumstances not foreseen by these presents.

ART. 13th

The Term for the existence of this association is limited to the first January 1810.

ART. 14th

Each of the concerned shall declare prior to the 1st January 1809 whether his intention is to withdraw from the association or to continue it.

ART. 15th

If the two-thirds of the concerned or more agree to renew the association, those who may wish to withdraw from it shall be reimbursed for the amount of their subscription & shares in the profits, as stated in the inventory of the 31st December 1809; and that payment shall be divided in three equal parts payable at three, six, & nine months from the date of their separation from the Company; each sum bearing interest at the rate of six per cent.

ART. 16th

If the two-thirds of the concerned don't agree to renew the association, E. I. Dupont will settle each and every account of the subscribers agreeable to the stipulations of the foregoing article.

ART. 17th

In either of the Cases provided for in Art. 15 & 16--the inventory of the 31 December 1809 shall be made by persons purposely appointed by the concerned parties, those persons to select an umpire, if necessary.

ART. 18th and Last
The present Deed of Association shall be considered as a legal title; and for that reason an authenticated copy of the same shall be delivered to each concerned.

Done in Paris the first floreal ninth year
21st April 1801

[Bidermann   Du Pont de Nemours
                                              Father, Sons & Co.

    Peter Bauduy   Catoire, Duquesnoy & Co.

                                            Arch^{d} McCall

                    E. I. DuPont   ]

   Longwood Manuscripts, Group 5, Box 49, December 31, 1809. This is a translation of the original Deed of Association in Duplanty's handwriting, apparently done at the end of the first partnership. The contents of the document have been altered to conform to current spelling, punctuation and capitalization norms for purposes of clarity. The substance of the original was maintained at all times. The information in brackets was crossed out on this document, although it appears in the original Deed of Association. The signatures at the bottom of the original document are valid, except that the firm DuPont de Nemours Father, Sons & Co. signing is in the handwriting of the unidentified drafter of the document. Accession 146, File 21.

Appendix B

ESTIMATED COST OF THE GUNPOWDER MANUFACTORY

Estimate of the costs necessary to establish
the Powder manufactory
----
to pay in the first year

Purchase of land about $6000
    of which one third is [downpayment].......... 2000
Machines made in France............................ 4000
Advances made to laborers coming from France....... 1000
Lodging for workers................................ 1000
Refinery... 1500
Wheel or pulverizing mill [for sulphur]............ 2000
Composition and charcoal house..................... 400
Stamping mill...................................... 2000
Graining mill...................................... 1200
Grainer and presses................................ 500
Glazing mill....................................... 1000
Dry house.. 400
Dust mill and packing house........................ 500
Magazine... 1000
Enclosure [fence].................................. 1500
House of the director [E. I.] and barn............. 3000

                                                                   ‾‾‾‾‾‾‾
                                                                   23000

    * [to pay in the second year and
    after]
        ----- Capital free for
    the purchase of raw materials, the construction
    of a second stamping mill, and the payments
    left on the land............................13000

                                                      ‾‾‾‾‾‾‾‾‾‾‾‾‾

                                                    36000

* The statement in brackets has been marked out on the original.

    Longwood Manuscripts, Group 3, #2367. This is a translation of the original document.

Appendix C

THE FIRST JOURNAL ENTRY

According to the first entries in this record book, the DuPont Company, as of April 21, 1801, was completely capitalized. Peter Bauduy recorded the capitalization of the firm in the first record book of the gunpowder manufactory as follows.

April 21, 1801

Brandywine   Powder Mills

Sundries to Stock $36000 amount subscribed for to the end of establishing a Gunpowder Mills in the United States of America for the Joint account of the Subscribers under the direction of E. I. DuPont de Nemours agreeable to the contract signed this day by the said subscribers viz.

14	Biderman's account proper	1 share	$2000	[1]
16	Duquesnois account proper	1 D^O	$2000	
20	Necker-Germany's account proper	1 D^O	$2000	
18	Dupont de Nemours, Father,			
	Sons, & Company	15 D^O	$30000	
1				36000

Stock to Sundries $36000 concern of the subscribers viz.

1

15	To Biderman's account in Co.	for his 1/18	$2000	[2]
17	Dusquenois D^O	D^O	D^O	
21	Necker-Germany D^O	D^O	D^O	
19	Dupont de Nemours, D^O	their 15/18	$30000	
	Father, Sons, & Company		36000	

Accession 500, Journal, #877, first written folio, not designated folio 1 by the bookkeeper.

[1] The Account Proper was the individual's account with the firm and included all transactions between the individual and the firm. Interest on invested capital as well as routine transactions were included in this account. The accounts of capital investors were the only ones annotated in this way.

[2] The Account in Company was the actual capital account for the investors in the firm. Only the capital transactions themselves such as the buying or transferring of shares were recorded here.

REFERENCES

Primary Sources

Hagley Museum and Library. Records of E. I. DuPont de Nemours & Co., 1801-1902. Accession 500, Series I, 1800-1818.

Hagley Museum and Library. Longwood Manuscripts.

Hagley Museum and Library. Winterthur Manuscripts.

Secondary Sources

Burchfield, R. W., ed. A Supplement to the Oxford English Dictionary. Oxford: Clarendon Oxford England Press, 1972.

Dilworth, Thomas. The Young Book-keeper's Assistant. 12th ed. Philadelphia: Benjamin Johnson, 1794.

Dufief, Nicolas Gouin. A New Universal and Pronouncing Dictionary of the French and English Languages. Philadelphia: T & G. Palmer, 1810.

DuPont, Bessie Gardner. Life of Eleuthère Irenée duPont from Contemporary Correspondence, 1778-1834. 12 vols. Translated and edited by B. G. DuPont. Newark, Delaware: University of Delaware Press, 1923-1926.

Edwards, Nina Lorraine. "The Bookkeeping Records and Methods of E. I. DuPont de Nemours and Company, 1801-1834." M.A. Thesis, University of Western Ontario, 1966.

Freedeman, Charles E. Joint-Stock Enterprise in France, Chapel Hill, N. C.: The University of North Carolina Press, 1979.

French, Sidney J. Torch and Crucible. Princeton: Princeton University Press, 1941.

Garner, John. Le Nouveau Dictionnaire Universel. Vol. 1: French--English. Rouen, France: Pierre Dumesnil et Fils, 1802.

Geijsbeek, John B., trans. <u>Ancient Double-Entry Bookkeeping. Lucas Pacioli's Treatise (A.D. 1494-the Earliest Known Writer on Bookkeeping) Reproduced and Translated</u>. Denver, Colorado: J. B. Geijsbeek, 1914.

Guerlac, Henry. <u>Antoine-Laurent Lavoisier</u>. New York: Charles Scribner's Sons, 1975.

Holland, Dorothy Garesche. <u>The Garesche, de Bauduy, and des Chapelles Families</u>. Saint Louis: Schneider Printing Company, 1963.

Institute of Chartered Accountants in England and Wales. <u>Historical Accounting Literature</u>. London: Mansell Information/Publishing Ltd., 1975.

L'Assemblée Nationale. "Loi Relative à la Fabrication & Vente des Poudres & Salpêtres." Paris: De l'Imprimerie Royal, 1791.

Mair, John. <u>Book-keeping Moderniz'd</u>. 6th ed. Edinburgh: Bell & Bradfute, 1793; reprint ed., New York: Arno Press, 1978.

Manuscripts Department. "Schedule of the Records of E. I. Du Pont de Nemours & Co. 1802-1902." Eleutherian Mills Historical Library, Greenville, Delaware, 1965.

McKie, Douglas. <u>Antoine Lavoisier; Scientist, Economist, Social Reformer</u>. New York: Henry Schuman, 1952.

Murray, James A. H., ed. <u>A New Oxford Dictionary on Historical Principles</u>. Oxford: Clarendon Press, 1888.

Riggs, John Beverley. <u>A Guide to the Manuscripts in the Eleutherian Mills Historical Library</u>. Greenville, Delaware: Eleutherian Mills Historical Library, 1970.

Saricks, Ambrose. <u>Pierre Samuel Du Pont de Nemours</u>. Lawrence, Kansas: The University of Kansas Press, 1965.

ANNOTATED BIBLIOGRAPHY

Primary Account Books.

The Blotters.

Hagley Museum and Library. E. I. DuPont de Nemours & Company Records. Accession 500, Series I, Blotters, 1800-1818.

Blotters were generally considered the books of original entry for transactions. Information relating to such transactions was recorded in these volumes in a rough form for later transfer to other account books. The emphasis was placed on adequately documenting the event. The correct entry form was added in other record books. Not all the volumes categorized as Blotters by the Manuscripts Department of the Eleutherian Mills Historical Library (now Hagley Library) were used for this purpose, however. Where other functions existed, the volumes have been identified accordingly.

1. No. 848. Blotter (Factory Building Book), 1800-1807.

The Factory Building Book was not a Blotter. Rather, it was supposed to be used as an expense book for the period of time during which the production facilities were being built. The record keepers continued to use this volume long after the production process was in operation, however. Use of the Factory Building Book as an active account book did not cease until December 31, 1807. The record keepers did not maintain the volume in a Debit and Credit format. They simply listed several individual expense items under the overriding heading "Manufactory Dr. to Cash for Sundries." The dates were not specific, nor was there a running balance maintained. The initial record keeper was Peter Bauduy. As of February 11, 1806, the bulk of the information amassed in the Factory Building Book was entered in the appropriate Journal by Raphael Duplanty. Accession 500, Blotter, #848, folio 52. Accession 500, Journal, #877, folio 71. He seemed to function as something of an internal auditor for this volume in that the individual entries included numerous annotations in his handwriting that indicated corrections or amendments to information contained therein. This leather covered volume was labelled "Ledger" on the binding. The volume was extensively repaired. The spine was rebacked with leather, with the original spine glued back over the new one. The label may have been part of the original spine, or it may have been added at the time of the repair. In either case, this volume was not used as a Ledger. The covers of this volume were completely re-attached. The original sewing was still intact, although some additional work was done on the stitching. The folio numbers remained consecutive, however, so it would appear that internal integrity was maintained despite the extensive repairs.

2. No. 849. Blotter (Day Book), 1804-1813.

Company record keepers did not use this volume as a Blotter, and it was not even one continuous record. Instead, the record book actually consisted of three completely different, unrelated records. The term Day Book, added by the Manuscripts Department of the Eleutherian Mills Historical Library (now Hagley Library), does not really seem to apply to any one of these. Both Peter Bauduy and Raphael Duplanty maintained this volume at various times from November 1800 to about April 1813. One document in this volume beginning November 20, 1800, was initially a duplicate of the Factory Building Book although less detailed. It was located at what appears to be the beginning of this record book, and it was maintained by Peter Bauduy. Beginning with September 15, 1802, the volume included only the entries from the Factory Building Book which were annotated with the initials "PB" for Peter Bauduy, apparently indicating items he had paid for that were subject to reimbursement by the firm. The last entry in this particular record was dated July 9, 1805. If the above document was located at the beginning of this volume, then in the back of the volume and upside down was another document, the first record of the company's powder sales. It was maintained by Peter Bauduy and only covered the period from the date of the first powder sale, May 16, 1804, to January 14, 1805. The primary function of the volume was documented in the middle of the book. These records were kept by Raphael Duplanty, and constituted the first payroll record, or "Petit Ledger," for the firm. There were individual payroll accounts including, generally, the dates of entry, employment, wages owed, time of employment, and sums paid out in cash or to others for: turnips, rum,

doctor's bill, pork, flour, extra boarding, and so on.
There were format changes from time to time, but
essentially the same information was documented.
Entries in the workers' accounts continued through
April 1813. The accounts in this volume were
continued in the next Petit Ledger. (Petit Ledger,
#961, not included in Annotated Bibliography) All of
the volume was used although in light of the limited
use of other volumes during the same time, it seems
unlikely that austerity measures required the three
uses of this book. This volume was extensively
repaired and rebound. There is every indication that
great care was taken at all stages to ensure that the
internal integrity of the documents was maintained.
There is therefore no reason to suspect that the
volume was not returned to its original state, with
three documents included in one book. The volume has
a decorative paper cover with a new leather spine and
leather corners.

3. No. 850. Blotter (Factory Book), 1804-1810.

        The primary purpose of this volume called the
Factory Book was to pick up the record keeping from
the Factory Building Book after the manufactory had
been built, and powder was being sold. It also
contained more than one document. The period of time
covered by this book was May 19, 1804 through June
1810. The first document in this volume was entered
under the title "Statement of the Different Sums Paid
by Irenée Dupont towards the Manufactory." Accession
500, Blotter, #850, folio 1. It began on May 19,
1804, and concluded on December 31, 1805. The
information indicating what happened, the date, and
the amounts involved was entered sequentially by Peter
Bauduy, and continued through folio 15. The total for
this period was Debited to the account of E. I. DuPont
in an entry in the Journal on February 11, 1806, and
the source record book was identified as the "Factory
Expenses Book." Accession 500, Journal, #877, folio
71. The next document in this volume was the actual
Factory Book and began on folio 16. The first entries
in the Factory Book for the balances from 12/31/06 and
12/31/07 indicated that the entries for this period
were "Paid in the Course of this year as per minute
a/c in factory building book thro' mistake."
Accession 500, Blotter, #850, folio 16. The real
entries in this document began on January 1, 1808, and
extended through the beginning of 1810. They were
basically one line entries, with the date, the nature
of the expense and the dollar value of the

transaction. The whole volume seemed to fall into disuse about June 1810. Information from the Factory Book was carried to two Journals for the period ending December 31, 1809. Accession 500, Journal, #877, folio 230. Accession 500, Journal, #878, folio 45. The information for the beginning of 1810 apparently was not included in the Journal. The Factory Book is in much the same condition as the Factory Building Book. Accession 500, Blotter, #848. This volume has a decorative paper cover with leather corners and a leather spine.

4. No. 851. Blotter, 1806-1808.

This volume was maintained by both Raphael Duplanty and Peter Bauduy, but primarily by the former. It covered the period from February 27, 1806 to December 31, 1808, and conforms to the definition of a Blotter. It therefore included a chronological listing of the events that affected the firm and was the book of original entry for this period. The individual entries were very rough, but they did identify the accounts and dollar values for each transaction. Separate entries were annotated with x's, apparently indicating that the bookkeepers followed some kind of an organized posting process. This information was probably posted to the appropriate Waste Book, which is no longer extant. On December 31, 1808, the x's disappear, and for the remaining few pages of the volume the entries were annotated with the Ledger folio numbers in the handwriting of Raphael Duplanty. Accession 500, Blotter, #851, folio 142. This was the process followed in the company Journals of this period. There were also annotations to the effect "Ent^d in Journal" in Duplanty's hand in these last pages, evidence of a sort of internal auditing function performed by Duplanty. Accession 500, Blotter, #851, folio 142. This book appears to be in its original condition. The original sewing is still intact. The volume has a decorative paper cover with leather corners and a leather spine. On the cover in the handwriting of the bookkeeper, Raphael Duplanty, is the notation "Blotter feb^y 27 1806 to Decr 1808."

5. No. 852. Blotter A, 1809-1810.

This volume was labeled Blotter A by the library. It is not clear why the A was added, although on the front of the volume, in the handwriting of Raphael

Duplanty, is the annotation "N° 2" and "1809 AA" with no additional explanation. This volume was maintained by both Raphael Duplanty and Peter Bauduy, although primarily by the latter. It covered the period from January 2, 1809 to December 31, 1810, and directly succeeded the previous Blotter. There were two documents in this volume. What passed initially for the front of the volume was actually a six-folio hodgepodge of summarized information on the firm's dealings with the U.S. government from 1811 through 1814. Unfortunately, the pages in the middle part of the book were cut although where they were attached at the spine remained intact. There is no way to tell if any information was lost, however. The primary document in the volume is in the back of the book from the entries mentioned above, and upside down. The volume has experienced a great deal of repair, but appears intact, nonetheless. The text block was reattached to the marble paper covers of the volume, with new leather on the spine. The corners of the covers were reinforced with leather. The spine was subsequently lettered "Blotter Day Book No 2" with the dates "January 2, 1809--December 31, 1810," annotated there as well. It is possible that the cover of this volume was reattached upside down. Nonetheless, the internal integrity of the documents has been maintained, and can be verified from the consecutive numbering of the existing pages as practiced by the company record keepers.

6. No. 853. Blotter, 1814-1817.

Record keepers used this volume as a straightforward Blotter. It was maintained by several different record keepers, although the only one that can be identified was Bauduy who started the volume. It covered the period from July 1, 1814 to June 30, 1817. There is no explanation for the gap in the Blotters from December 31, 1810 to July 1, 1814, although the intervening period may have been covered by a Blotter that is no longer extant and not mentioned in the other record books. This volume is covered in marble paper. It has leather corners, and a leather spine. The original sewing is still intact and the volume shows no evidence of repair.

7. No. 854. Blotter, 1817-1822.

This volume covered the period from June 2, 1817 to December 30, 1822, although record keepers used it

as a Blotter for only part of this period. It was
maintained by several different record keepers,
although the only one who could be identified was
Raphael Duplanty. Parts of the first and second
folios appear to be a continuation of the previous
Blotter. Beginning with the first folio one
occasionally sees the notation "PL" before the entry.
This denotes the Petit Ledger, and the information can
be traced to the appropriate Petit Ledger folio
number. Over time these entries appeared to be more
and more prevalent and eventually the volume became
primarily a support document for the Petit Ledgers
although there were, occasionally, Waste Book
notations. Duplanty again performed an internal audit
function. The volume is in its original condition.
The covers are marble paper, with leather corners and
a leather spine. There has been no repair work.

The Waste Books.

Hagley Museum and Library. E. I. DuPont de Nemours &
    Company Records. Accession 500, Series I, Waste Books,
    1810-1818.

    The Waste Books were the books of original entry for
transactions if no Blotter existed. The entries in these
volumes were more formalized than the Blotter entries and
the information was posted from the Waste Book to the
appropriate Journal. One Waste Book is no longer
extant. All volumes in this category for this period were
used as Waste Books, and referred to as such in other
account books.

1. No. 862. Waste Book B. 1810-1816.

        The first volume in the library's category Waste
    Books was labeled B. There was a previous volume,
    presumably Waste Book A, referred to in the Journal
    which is no longer extant. Accession 500, Journal,
    #878, folio 13. This volume was maintained by many
    different record keepers, including Duplanty. It
    covered the period January 1, 1810 to October 31,
    1816. The beginning date was the date when a new
    Ledger was started. Accession 500, Ledger, #935.
    This volume was organized chronologically and followed
    a general entry format. The entries in the Waste Book
    were checked, presumably to indicate a posting
    notation when the information was transferred to the
    appropriate Journal. The very beginning of this

volume included a detailed Inventory of all amounts owned offset by amounts owed and invested including cash, property, receivables, payables and all of the material goods of the firm. This volume is leather covered. It has experienced little repair work. Apparently the text block may have come loose from the leather spine, and may even have completely fallen out of the cover. This has been replaced inside the original covers which have been reinforced inside. The integrity of the document has been maintained, however, as the original sewing is still intact. The corners of the covers and the head of the spine were reinforced with leather.

2.   No. 863. Waste Book C, 1816-1818.

The second volume covered the period November 2, 1816 to August 31, 1818, and directly succeeded the previous Waste Book. It was used for the same purpose as the previous Waste Book. This volume was maintained by many different record keepers, including Duplanty. It is leather covered. The head and tail of the spine have been repaired. The text block has been reinforced at the inner joint and has been glued to the spine. The cover has been reinforced at the corners more recently. The integrity of the account book has been maintained.

The Journals.

Hagley Museum and Library. E. I. DuPont de Nemours & Company Records. Accession 500, Series I, Journals, 1801-1818.

The Journals included the same information found in the Waste Books and Blotters, although the chronological listing of the entries was not always maintained. Accounts included in the Journal entries were annotated with the appropriate Ledger folio numbers. All four volumes in this category qualify as Journals.

1.   No. 877. Journal A (Day Book), 1801-1808.

This was the first Journal for the firm. The book covered the period from the date the original partnership was formed, April 21, 1801, to December 31, 1808. The Journal was maintained by Bauduy until February 11, 1806, after which Duplanty made some of the entries. Each entry in this volume was annotated with the Ledger folio number in the handwriting of Raphael Duplanty. The book covers and spine are covered in leather, but the volume is in generally poor shape. No repair work was done on this record book. The front cover and several pages are off completely and most of the remaining text block is almost completely severed from the rest of the volume. One blank section at the back of the volume is completely out as well. The volume's integrity is maintained, however, because each page is sequentially numbered and there is no inordinate break in the sequence of dates. The pages, though somewhat stained, tattered and faded are all legible.

2. No. 878. Journal B (Sales Book), 1809-1810.

This volume directly succeeded the previous Journal. It was maintained by Duplanty and covered the period January 1, 1809 to December 31, 1810. This period overlaps the time frame covered by the succeeding volume. The purpose of this duplication of information cannot be determined. Only about one-third of the volume was used to record the information. The volume has experienced no repairs. It is leather covered. The original sewing is loose, but still intact and the volume has retained its integrity.

3. No. 879. Journal "B," 1810-1816.

This was the third Journal, although it was labelled with a B. The volume was maintained by Duplanty and by other record keepers who cannot be identified. This account book covered the period from January 1, 1810 to May 31, 1816. It covered the same period in 1810 as did the previous Journal although why the information was included in both account books is not clear. This book is leather covered and has been repaired. The spine was rebacked with leather, with the original spine glued back over the new one. The text block has been completely reattached. The individual sections were resewn, but were still consecutive, so one can assume that the book has retained its integrity. Leather corners have been used to reinforce the book on the front and back. The entire volume has been used.

4. No. 880. Journal "C," 1816-1818.

The volume was maintained by Raphael Duplanty and by other record keepers who cannot be identified. It covered the period May 31, 1816 to August 31, 1818, and directly succeeded the previous Journal. The record keepers deviated from the standard day-to-day entry format in January 1818, however. They seemed to be trying to organize the material in a different way, possibly so that the posting process to the Ledger would be facilitated. At first, the Debit part of each entry was included under an overall heading and totaled. The same thing was then done to the Credit entries. Initially, efforts were made to identify the debits or credits in different transactions for a particular account, and post them at one time. Over time the rigor of this practice diminished. Toward the end of the volume a few pages of such information were included at a time, so that there were many groups of Debit or Credit entries for the same account. This volume is leather covered. The text block has been completely reattached. The spine has been rebacked with leather, with the original spine glued back over the new one. The individual sections were resewn, but were still consecutive. The book has retained its integrity. Vellum corners have been used to reinforce the corners on the front and back.

The Ledgers.

Hagley Museum and Library. E. I. DuPont de Nemours & Company Records. Accession 500, Series I, Ledgers, 1801-1818.

The account information in each Journal entry was posted to the appropriate folio location in the Ledger. The significance of the debit or credit balance in each account after all entry information was posted depended on the nature of the account. The account balances accumulated in the Ledger were occasionally closed, and used to prepare Trial Balances. The Trial Balance served to identify the Profit or Loss for a particular period. The three volumes in this category qualify as Ledgers.

1. No. 934. Ledger "A," 1801-1810 with separate index.

This was the first Ledger. The volume was maintained by Raphael Duplanty and covered the period April 21, 1801 to 1810. In all likelihood Duplanty started this volume after completing the internal audit he performed in February 1806. The official ending date for this Ledger was December 31, 1809. The books were officially "closed" on this date, and a Trial Balance listing all Debit balance accounts first and totals, and then all Credit balance accounts and totals was drafted. Accession 500, Ledger, #934, folio 224. Certain 1810 entries were nevertheless included in this volume. This volume was maintained so that the left hand page constituted the Debit page, and the right hand page constituted the Credit page. Each entry was annotated with both the Journal folio number where the original entry was located, and the corresponding Ledger folio number(s) where the offsetting account information could be found. This is a leather covered account book which is in remarkably good condition. There has been some repair work on the volume, but the integrity of the material has been retained. The text block has been reattached and the spine has been rebacked with leather.

2.   No. 935. Ledger "B," 1810-1816.

This volume directly succeeded the previous Ledger and was maintained by Duplanty and several other unidentified record keepers. It covers the period from January 1, 1810 to December 31, 1816. Where 1810 information was recorded in the previous Ledger, no duplication occurred. The accounts that stayed open in the preceding Ledger were simply posted to this Ledger at a later date. In this volume Debits and Credits were recorded on one folio divided by a line drawn down the center of the page. This volume included two Trial Balances. The first of these was dated June 30, 1814. Accession 500, Ledger, #935, folio 211. The second one was dated December 31, 1815. Accession 500, Ledger, #935, folio 213. There was no Trial Balance when the ledger was closed on December 31, 1816. Instead, each account was simply closed to its new account in the succeeding Ledger, but the information was not compiled anywhere. This volume is leather covered. The text block has been resewn. The corners were repaired and the spine has been rebacked with leather, but the integrity of the material has been retained.

3.   No. 936. Ledger "C," 1816-1818.

This volume directly succeeded the preceding Ledger, and was maintained by Duplanty and a number of other unidentified record keepers. It covered the period from January 1, 1816 to August 31, 1818. The account format remained the same. The volume has undergone extensive repair. The text block was reattached. The spine was rebacked with leather and protective vellum corners were attached to the covers. Although the sections of the text block were resewn, the integrity of the volume was maintained.

Secondary Account Books.

The Cash Books.

Hagley Museum and Library. E. I. DuPont de Nemours & Company Records. Accession 500, Series I, Cash Books.

The Cash Books included information on cash disbursements and outgoing checks only. No receipts of cash were documented in these volumes. Two volumes cover the period under consideration, but as these are not primary record books for the DuPont Company, only one representative volume is included here.

1. No. 1035. Cash Book, 1810-1813.

This volume was maintained by Raphael Duplanty. It covered the period from January 1, 1810 to December 31, 1813. Each page includes cash or check information under several columnar headings. Thus, there is a category for amounts paid in cash, amounts charged in accounts current, and others for the accounts of E.I., Peter Bauduy, the woolen factory called DuPont, Bauduy & Company, and for the gunpowder factory. These amounts were broken down between wages and bills. These columns changed somewhat during the period in question to reflect changes in the firm. The information in this Cash Book was posted to the appropriate Journal at the end of the month. This volume was actually four, separate, paper-covered booklets bound together with a plain paper cover. Raphael Duplanty's handwriting appears on the front and the spine of the bound volume which was marked on the spine "CB A" and "Cash Book A Jany 1, 1810 to Decr 31st 1813," and on the front "Cash Book Jany 1st 1810 to Decr 31st 1813." It does not seem to have been handled much after the booklets were bound.

The Accounts Current.

Hagley Museum and Library. E. I. DuPont de Nemours & Company Records. Accession 500, Series I, Accounts Current.

The Accounts Current volumes contained information on very active accounts, as well as serving in some cases as Memorandum Books. Several volumes qualify as Accounts Current. The first of these volumes is considered representative of this general category of subsidiary record book.

1. No. 1064. Accounts Current (Cash Book), 1804-1808.

This volume contains separate accounts bearing the names of some of the individuals who were prominent agents of the firm during this period. The handwriting in this volume belonged primarily to Raphael Duplanty. This volume covered the period December 20, 1804 to sometime in 1808. It was difficult to interpret, primarily because it included a veritable hodgepodge of different items. A lot of the interpretation that goes along with this volume seems to depend on which folio and account within the volume one is dealing with at the time. In keeping with the primary purpose of this volume, it included a detailed account of the state of exactly what the firm owed to certain individuals and what those same individuals owed to the firm. In most cases the amounts and configuration differed from the Ledger, but many individual items may be traced to the Journal and from there to the Ledger. There is no formal annotation to determine this, except for correlation of the dated transactions in the Accounts Current with the same information either at the same date, or some close date, in the appropriate primary account books. Within this volume references were made to individual Accounts Current which were sent by the various agents. This coincided with the information in the letters which constantly referred to the need to receive these accounts, or reports, from the various agents throughout the country. There was also a reference in this volume to efforts to reconcile the information received from the various agents with the information maintained by the firm. One such entry stipulated that there was an "Error which it appears we have committed in making the above extracts from their accounts which makes our a/c agree with theirs." Accession 500, Accounts Current, #1064, folio 8. This volume also seems to have been used to maintain the Accounts Current of the various shareholders in the firm. Additionally, there were references to certain firms with which the principals in this firm had close ties. These included Bauduy, Garesches and Company, and Dupont, Bauduy and Company. This volume appears to be intact, but its condition is not good. The textblock is not sound. Most of the sewing is gone, and of the original cords, very few remain. There are, however, no evident gaps in the material. The existing pages are numbered sequentially except for the first, unnumbered page. There is some danger with this volume, however, that parts of it may be lost, however, since the front cover is completely loose.

ACCOUNTING HISTORY AND THOUGHT

An Empirical Study of Financial Disclosure by
Swedish Companies.
> T. E. Cooke

Accountability of Local Authorities in
England and Wales, 1831–1935.
> Edited by Malcolm Coombs and J. R. Edwards

Accounting Methodology and the
Work of R. J. Chambers
> Michael Gaffikin

Schmalenbach's *Dynamic Accounting* and
Price-Level Adjustments.
An Economic Consequences Explanation
> O. Finley Graves, Graeme Dean, and Frank Clarke

An Analysis of the Early Record Keeping in the
Du Pont Company, 1800–1818.
> Roxanne Therese Johnson

The Closure of the Accounting Profession
> Edited by T. A. Lee

Shareholder Use and Understanding of
Financial Information
> T. A. Lee and D. P. Tweedie

The Selected Writings of Maurice Moonitz
> Maurice Moonitz

Methodology and Method in History
A Bibliography
> Edited by Lee D. Parker and O. Finley Graves

Accounting in Australia
Historical Essays
> Edited by Robert H. Parker

*The Growth of Arthur Andersen & Co., 1928–1973
Leonard Spacek

The Cash Recovery Rate Approach to the Estimation of Economic Performance
Edited by Andrew W. Stark

*Studies in Accounting Theory
Edited by W. T. Baxter

The Story of the Firm, 1864–1964,
Clarkson Gordon & Co.

A Half-Century of Accounting, 1899–1949
The Story of F. W. Lafrentz & Co.

Touche Ross
A Biography
Theodore Swanson

The District Auditor
Leonard Mervyn Helmore

*Academy of Accounting Historians Classics Series

Garland publishes books on all aspects of the accounting profession; for a complete list of titles please contact the publisher.